The Challenges of Time

Walter Grassi

The Challenges of Time

Myth, Physics, Environment

Walter Grassi
Department of Energy, Process and System
Engineering
University of Pisa
Pisa, Italy

ISBN 978-3-030-94371-4   ISBN 978-3-030-94372-1   (eBook)
https://doi.org/10.1007/978-3-030-94372-1

This Springer imprint is published by the registered company Springer Nature Switzerland AG
The registered company address is: Gewerbestrasse 11, 6330 Cham, Switzerland

To my wife Letizia.
You will meet her, she who is not a globetrotter, on the trains that I will need to
exemplify a number of concepts (Jacopo is my nephew, too small to travel alone).
You will see her occasionally reading a book. I don't say it in the text, but she's so
masochistic and I'm so sadistic that it's about the drafts of mine.
We've been traveling together in good and bad weather our whole lives.

Preface

Writing a text is never easy. For those who are accustomed, as I am, to deal with specific topics and want to make an attempt to divulge some of the things, they have learned (even if only on the sidelines of their own field of competence) it is prejudicial to define the objective they have in mind. Mine is very modestly to intrigue the reader, pushing him to be interested in various subjects that I think are interconnected and to highlight as much as I can such interconnections. Precisely because I am a specialist, I know the danger that is surreptitiously hidden in excessive specialization (however necessary) and I have always considered it fundamental to privilege the method over the particular notion. Moreover, I am an advocate of the vision of a complex reality and of the need to tackle problems with an approach that takes its cue from the science of complexity. This science teaches us that it is neither single individuals nor single events, alone, that determine the peculiarities of a system, but they are deeply connected by their reciprocal interactions. Just as often, it is not the single particular points of view that allow a complete knowledge of a concept. They can define some characteristics of it and thus delineate it better. It is, however, probable that the points of view with which it is looked at change with time and it is, in any case, difficult to grasp all the different facets and, even more, to make a synthesis.

On the north side of the Cathedral of Pisa in a marble of Roman origin are many small holes, which they call the "fingers of the devil". Legend has it that the devil climbed up the side to prevent the construction, but was overwhelmed by the divine power. The holes remain as evidence of his claws in the climb. Each time, you count them they give you a different result. It is a "frustrating" experience for those who repeat the count hoping for a result; the number of tourists who do the experience is also great. You continue, each trying to find out the number or at least to get an approximate idea. Even with the concept of time, I have not done, here, much more than this.

Pisa, Italy   Walter Grassi

Contents

1

Introduction

We are so used to using certain words that we do not ask ourselves what they mean. This is the case with the word time, which marks our lives by dividing it into past, present and future. We are so familiar with it and measure it so often, with watches, mobile phones, with the passing of days, seasons and years that it has become an integral part of our lives. We even say goodbye to the past year and welcome the next one, without perhaps realizing that, in doing so, we try to exorcise its inexorable passing by clinging to a sort of tranquilizing cyclical nature. With our time, we also measure that of everything around us, treating it as absolute time. But is it really so?

The science of the last two centuries has given a strong shake to this conception, thanks to the fortunate combination of brilliant thinkers and a technological progress that has allowed the experimental verification of their thought.

"Unfortunately", this concept we are used to is not as absolute and universal as we have always thought. Perhaps, the best known contribution, or at least the most popular, is that of Lorentz, Minkowski and Einstein, who took away from physicists, and not only from them, this illusion, essentially born with Galileo and Newton. Between the end of the nineteenth and the beginning of the twentieth century, they supported the idea that time should be considered only as a fourth dimension to be added to the three spatial ones. And they seriously doubted its autonomous existence with the characteristics we have mentioned. All this was inspired by Maxwell's theory on

© The Author(s), under exclusive license to Springer Nature
Switzerland AG 2022
W. Grassi, *The Challenges of Time*,
https://doi.org/10.1007/978-3-030-94372-1_1

the propagation of electromagnetic waves and was manifested by the experimental study of phenomena (from radio waves to elementary particles) that occur at speeds close to the speed of light. Very fast phenomena therefore, happen in extremely short times.

In the same period, Wegener (1880–1930), with his theory of continental drift, described the earth's changes on a temporal basis of orders of magnitude infinitely superior to the duration of animal life. Darwin himself, with the publication of the "Origin of Species" in 1859, proposed a revolutionary explanation of why there is such a great variety of forms of living beings. In this way, he highlighted how the evolutionary process had to be prolonged for a very long time, much longer than the dates initially referred to for creation. Time thus marked the stages of the earth's development both in relation to its geological morphology and in relation to the production and evolution of life. On a much more limited scale, it also measured the existence of individual living beings, generally unable to perceive the changes that took place over much longer periods of time.

Whether they referred to very short times or to very long periods (evaluated with our timescale), these theories found more or less strong oppositions, that, in any case, forced to great debates and, finally, strengthened them. It is enough to remember the long-standing discussion on the existence of the ether and the adversities encountered by the theory of relativity. And think of Darwin, who also drew harsh criticism even from the church. Moreover, the debate between creationists and non-creationists is still very lively in some countries, particularly in the USA. Continental drift did not have an easy life either. This theory was initially very badly received by the scientific world. Some even went so far as to call it a fairy tale. Even in the 1950s, the theory was only mentioned in university teaching for the sake of the record. It was geophysics and stratigraphy together with the evolution of experimental technology that established its veracity. Some date back to a 1960 communication to the Royal Society of London the confirmation of the differential displacement of the continents.

In 1800, a new science, thermodynamics, began to take shape and continued into the next century. With it, they learned to distinguish between systems made up of a large number of individuals and a few isolated individuals. For the former, a time appeared with clarity that marked the phenomena and accompanied them until their exhaustion. Due to the effect of irreversibilities, its direction was univocal and people began to speak of the arrow of time. Subsequently, it was understood that time was not an "external spectator" of what was happening, but assiduously participated in making it happen. Thus, the theories of dissipative structures and complexity were developed.

All this does not make it clear to most of us which time we should believe in. All the more so because it seems to be well established that the passage of time is perceived differently by different living organisms. For the same individual, this perception also changes according to the contingencies and even according to the metabolism. This is why, for example, as we age, everything seems to pass more quickly. Or happy periods seem to pass in a flash, while difficult ones never seem to pass.

There is, however, an aspect that forces us to take up the concepts expressed by thermodynamics, that is, those of time with a preferential direction and builder, or destroyer, of phenomena: the environment. In it, we speak of large numbers, of cause and effect and of different times for the planet earth and the beings it hosts. Here, the game becomes heavy and the survival of the entire system or, as Lovelock called it with a suggestive vision, of Gaia, depends on it.

These are the topics that we will deal with in the following in a necessarily synthetic way, simplifying and exemplifying as much as possible the concepts exposed.

Finally, the reader is warned that from time to time in the text are included mathematical insights in special boxes. Those who are not interested can skip them without any inconvenience for a coherent reading of the arguments.

2

From Myth to Experimental Science

2.1 Walking My Dog

Every day, I go out with my dog (Fig. 2.1) for walks of a few kilometres in the meadows. His name is Iago, he is a malamute dog of the same size as the German shepherd, and in the meantime, he does many more kilometres than me going up and down without getting tired. He looks like he could do about forty a day attached to a loaded sled. He is always with his nose to the ground smelling smells that I do not even know exist. He hears sounds that I cannot hear and sees things differently and at a higher speed than I do. His hearing perceives sounds coming from distances four times greater than those from which we perceive them, and the range of frequencies within which he hears sounds is about double that of ours, being able to pick up ultrasound. If he watched television, he would be able to separate the individual frames and would not see, as I do, a fluid succession of events [1]. In addition, the spectrum of colours he sees is essentially limited to yellow, blue and shades thereof.

Not to mention the sense of smell, a sense to which in the dog are dedicated a number of papillae and an olfactory cortex enormously larger than those of man.

We are so different, and yet, we "get along". No wonder my dog is smart, and I could answer you as "master", clumsily measuring his intelligence with the ability to adapt to mine. So also scientists, to study the sensory and perceptive capacities of dogs, do not have an objective method and are forced

© The Author(s), under exclusive license to Springer Nature
Switzerland AG 2022
W. Grassi, *The Challenges of Time*,
https://doi.org/10.1007/978-3-030-94372-1_2

Fig. 2.1 My dog Iago

to compare them with ours. That is, they look at the object of their investigation with the glasses of the investigating subject. They are, however, aware of this and seek to objectify their conclusions through the application of a scientific method of repeatable experimentation.

The common "master" is obviously not a scientist, and his dog's understanding is, by necessity, influenced by the narrowness of his knowledge and often by the notion that man, made, for some, in the likeness of God, has no need to understand, but rather "deigns" to do so. In fact, since we live in two partially different worlds, with different priorities and sensory references, we must establish between us a conventional way of interacting. This is based on vocal and gestural expressions, which over time and with patience become our common language. If we do not do this and everyone lives in his own restricted environment, the only unfortunate possibility is constriction on the part of man and fear and sometimes violent reactions on the part of the dog.

The only serious possibility is to create a true and profitable interaction, which allows both to understand each other by establishing clear rules and not only due to an instinct of mutual overpowering and fear. It is necessary, as with all things, to be able to maintain an attitude of serene openness and desire to know what we have around us, without strange fears and/or prejudices. First try to know and understand without relying too much on what our often superficial observations tell us.

As I walk, I think about what it would be like if I had different sensory abilities. I could hear the sound of growing grass, ants walking through the grass, small mammals in their burrows and so on. Certainly, I would have a fuller perception of the life that flows around me parallel to my own, feeling much more a part of something bigger, but, probably, rather confused in a sort of cacophony. Perhaps, it is our sensory limitations that allow us to live in a "coherent" way, but it is fair to acknowledge that they are still limitations.

As I watch Iago, it occurs to me that it is said that for dogs every year is worth seven of ours and that they live in the present. I think it means they do not make plans for the future and maybe they do not even know what that means. On the other hand, we too can only live in the present by remembering the past and making predictions and plans for the future, of which we have no certainty.

Then, it comes to mind to think about what the present is and even what time is. The curiosity to look around you is great, and you know that you are certainly not the first to have thoughts of this kind and not even "equipped" to enter the philosophical meanderings of this topic. You can only limit yourself to some reflections as an ordinary man and to the hints that your cultural background can provide. For example, I realize that I have a time of my own that began when I was born and will end with me. It is connected with a historical time that conventionally places mine in a sort of "universal" time, however, marked by man. A beautiful sentence of Shakespeare's in Macbeth comes to mind: "*Life is but a walking shadow; a poor actor that struts and frets his hour upon the stage and then is heard no more: it is a tale told by an idiot, full of clamour and fury, signifying nothing*". (W. Shakespeare, *Macbeth*, V, 5, 23–28).

But I also have my own "timing" in doing things just as I have my own way of using time and organizing it.

I think this is a subjective time as it is influenced by our perception.

Moreover, we are still accustomed to think that there is an objective time that flows autonomously and independently from the perceived. Yet, Einstein made his revolution with the theory of relativity more than a century ago, but some ideas take a long time to enter the common mentality, also because of their complexity. Or maybe it is all due to the fact that Einstein did not have Twitter, now for many a unique tool to communicate "thought", on whose depth and awareness it is better not to investigate. But nowadays, life is convulsive and there is no time: first you "twit" and then, perhaps, you think.

Curiosity pushes me to go far away when man thought of the earth as the centre of the universe and still felt part of nature. All that remains for me is to poke around in a world that is at least partially unknown, trying to understand, if nothing else, the most established things in scientific thought and beyond.

2.2   From Kronos to Aristotle

The ancient Greeks tell that, at the beginning, nothing existed in creation, only the formless chaos beyond space and time. Suddenly Gaea, the earth, mother of creation, appeared. First, she generated Uranus, the sky, who in turn showered the earth with a beneficial and fertile rain and then Pontus, the sea. She then married Uranus, and together, they governed creation and had several children. Kronos was one of them, but with a particular destiny. In fact, for fear of being deprived of his dominion over the universe, Uranus made him sink into the bowels of the earth together with his brothers. Kronos, however, instigated by his mother, confronted him and emasculated him. He then joined in marriage with his sister Rhea, to continue the process of creation. From this union were born several children and his kingdom prospered, until he was prophesied that it would end by one of them. From this moment, Kronos began to devour the newborns, keeping them prisoners in his bowels. The only one who was saved was Zeus, thanks to a stratagem of Rhea. The story goes on, but we stop here to try to give a meaning to it all. According to Rindone [2], the emasculation of the father by Kronos means a "*depowering of the past by wrapping it in oblivion*" and, devouring the children, "*destroys the future as soon as it comes into existence*".

Using Rindone's words again, let us try to summarize the vision of the Greeks. "… *For the Greeks, in general, one cannot attribute meaning to the adventure of humanity on earth. For them, in fact, man, whose excellence they exalt, does not emerge radically from nature, the original womb from which everything is produced and in which everything returns to dissolve… Since human life is part of the cyclical becoming of nature, the most common conception of history is precisely the cyclical one: even the great works of man, of which we keep memory, are destined to perish, because everything is born and dies, everything returns and it would be vain to imagine radical changes. There is no escape from cyclical becoming and its necessity, except with the tension towards a transcendent and eternal reality. This world must therefore be accepted as it is: the most significant exception, as we have seen, is Plato…*".

Everything flows (Panta Rhei) said Heraclitus, and everything is never exactly the same as itself in the previous instant. Classic is the sentence that says that you cannot bathe twice in the same river, that is, in the same water flowing in the river. It is already the vision of a dynamic reality, which changes continuously, to which man must adapt or rather that he must undergo. And, since he is part of this reality, he too is subject to continuous change. All this provokes a sense of the ephemeral and the uncertain which disorients. This is why man turns to the supernatural: a stable and unchanging reality, free from change, populated by gods and heroes. Using Rindone's words again: *"By repeating in the gestures of daily life the exemplary behaviour of the mythical hero, the primitive thus escapes the senseless succession of time and its chaotic randomness, living in a divine present rich in meaning"*.

Both the transcendent and eternal reality, which establishes a sort of continuity between human and divine history, and cyclical time, are elements that synergistically seem to want to give meaning and stability to a reality that otherwise lacks it.

Just as the sun rises and sets every day indefinitely and the seasons repeat themselves, man performs the same actions over and over again, following the rhythms of nature, of which he is well aware that he is a part. Even generations follow one another regularly and even history repeats itself cyclically. This eternal return of things to infinity and in the same state contrasts with the constant becoming, since every event repeats itself periodically. Time is represented as a wheel, and everything repeats incessantly, for some in the same way and for others with some variation. It was not only the Greeks (especially the Stoics) who supported the circularity of time. It was present in some eastern religions and was also taken up much later by Nietzsche himself (1844–1900).

In this way, man tries to annul what today we would call the non-reversibility of time, that is, the impossibility for time to go backwards. If we think about it, we also carry out the same gestures and do more or less the same things every day. We scan time by dividing it into hours, days, months and years and we identify certain holidays during the year in a "reassuring repetitiveness". We say goodbye to the past of the old year and welcome the future of the new year with a kind of propitiatory rite.

But let us dwell on a few thinkers.

Aristotle can be a good starting point for the discussion we want to have about time because of the profound influence that his thought has also exerted on Western scientific culture. It is enough to remember the locution "Ipse dixit" (today we would say: he said it) which, although initially referred to Pythagoras by the Pythagoreans, as Cicero recalls, from the Middle Ages calls into question Aristotle as the "highest authority". In fact, the rediscovery of Aristotle and Greek thought began in the West around the twelfth century. It was present in the Latin world, but ended up disappearing with the fall of the Western Roman Empire (486 A.D.). The first to take possession of this heritage were the Arabs from the middle of the eighth century, translating it into their own language and fusing it with their own culture. In the meantime, these people had made many advances in various fields of science, including mathematics and astronomy, showing great interest in pure science.

Latin retranslations began in Italy about four centuries later and spread to Spain.

The "fortune" of Aristotelian philosophy (initially opposed by Christian thinkers) in the following centuries was due in particular to its progressive fusion with Christian theology, so much so that it took over from Plato's mystical vision.

Aristotelian physics survived for centuries and the major protagonists of the seventeenth century revolution had to "fight" against it. However, it continued to be taught perhaps even afterwards and constituted the main "obstacle" to the creation of present-day physics. This is the problem that arises whenever, instead of looking at knowledge with an open mind and willing to recognize its dynamism and, therefore, also its errors, it is accepted as a true and proper verb or, if you prefer, as a revealed truth.

In this way, the knowledge acquired does not constitute a springboard towards new and better goals, but instead becomes a real preconception, perhaps reinforced for political or religious reasons.

The first of the Western scientists to oppose it was probably Galileo Galilei (1564–1642) with the discovery of lunar craters, in contrast to the Aristotelian definition of a smooth and incorruptible moon and a perfect celestial world. The line-up of opponents then includes Descartes (1596–1650), Newton (1642–1727) and several others.

But let us see who he was.

Aristotle (384 or 383–322 B.C.), son of Nicomachus, personal physician to the Macedonian king, Aminta, joined the Platonic academy in Athens at the age of eighteen, where he remained until the death of Plato, his teacher, some twenty years later.

There were many differences between the thought of the master and that of the pupil. The latter maintained his critical sense, reworking through his own convictions and abilities the ideas of the master, if necessary, opposing them. The master, for his part, fully respected the freedom of thought and the cultural autonomy of the disciple. Of course, we are talking about great thinkers here, nothing to do with those many "little men" who intend to impose their positions (calling it thought is a bit too much) with the arrogance of empty and superficial slogans on the not always blameless common people.

Plato defined time as "the moving image of eternity that proceeds according to number". Time is a measure of the movement of the material world alone in which past, present and future have meaning. As if he thought of movement as the passing of hours and days. All this, however, lives in the sensible world where knowledge derives solely from the understanding of sensible phenomena, which are subjective when not contradictory and falla-cious. Therefore, it is the world of "opinion" (doxa) or even of appearance. Truth, on the other hand, lies in the knowledge of supreme concepts that are not bound to time and the material world, but to eternity, in which time has no place, and to the immutability of the world of ideas: the hyperuranium. This aspect is best clarified by the myth of the cave in which men, chained, take the shadows, which are opinion or appearance, for the truth of universal concepts. He does not, however, reject opinion since it represents a first step on the road that leads to truth.

For Aristotle "opinion" has instead its own intrinsic validity as an indis-pensable premise of scientific knowledge. There has always been and always will be time and, with it, movement or becoming.

Also with this assertion he manifests how the concept of eternity is funda-mental in his thought. This is in contrast with what the Christian conception of creation will say, therefore of a beginning and an end of the world.

There is a principle from which the world develops, the Prime (unmoved) Mover, which is the first cause of the evolution of the universe and from which logically space and time also derive, two absolute entities, but connected. Mover because it is the cause of everything and that towards which everything tends, immobile because it is the cause of everything, but it is not caused by anything because it already exists as a "pure act". Aris-totle thinks that becoming is a passage from potency, and let us say with today's language potentiality, to the act, the realization of this potentiality, which presupposes the existence of a being in act. The seed (potency) that germinates presupposes the existence of a plant. Only the existence of the plant (being in act) determines the evolution of the seed. Matter evolves in this way. God, instead, as a pure act is eternal and in him there is no matter whatsoever that is, on the contrary, in continuous transformation and corrup-tion. The world in which the "movements[1]" take place (the word movement is associated both to the real movement and to the becoming of things as they

---

[1] Natural beings obey the principle of change, which is an inherent characteristic of them. In particular, Aristotle indicates four types of change.

The first, considered a priority, is the change of position or place, which we still refer to as movement. It can be local or of translation, natural or violent. To this are added:

- The radical change of the substance of things, which are generated and then progressively corrupted. Birth and death are perhaps the prime example of this change. Only this kind of change implies that there is corruption of substance.
- The qualitative change of the substance. Summer tanning is an example.
- The change associated with quantity. A growing child, plant or animal is a typical reference of increasing quantity. Just as the water contained in a glass from which one drinks is an example of a decrease in quantity.

change), is divided into two spheres one celestial, or supra lunar, in which the only possible movement is the circular one that has no beginning and no end and a sublunar one. Here is eternity again. The very motion of the planets is circular, and in this it only imitates the eternity of God. Celestial space is composed of an incorruptible matter different from that of the material world, called "ether". We will find this word again much later, initially used to explain the propagation of electromagnetic waves.

The other, lower sphere, where material life takes place, is the sublunar sphere. All kinds of movement take place in this sphere, especially recti-linear movement. Matter is made up of the four sublunar elements and their mixtures: earth and water, heavy elements that spontaneously move down-wards, and air and fire, light elements that move upwards. The bottom is the position towards which heavy bodies naturally tend, and the top the position towards which light bodies go.

What about space and time? We have already said that they are absolute, and let us see in what sense. First of all, it means that they are equal for all and immobile. Time is therefore separate from the events that happen in it and from the beings to whom they happen, just as space is something separate and distinct from the bodies that are contained in it, or even of which it is the "place". In fact, the same region of space can contain different bodies. The part of space delimited by a thermos can contain coffee or tea or just air, but it has nothing in common with any of these substances and maintains its own autonomous existence. Space is also continuous. Between two points in space there is still only space and the point, as such, is isolable only virtually. As we would say today it is conceivable only as a limiting concept. The continuity of space is continuity in extension. If we refer to a segment of a given length it is always divisible into smaller segments.

In the space those movements (and therefore their becoming) take place in the direction of the natural positions towards which matter tends in the absence of obstacles, positions that are objective and do not depend on the observer. The sublunar sphere, of finite extension like the celestial sphere, is characterized by a high and a low and is the seat of translatory motions of substances towards the natural positions.

The celestial sphere with which the sublunar one adjoins does not admit these motions of translation, but only circular ones, being perfect and eternal in itself.

Even the sublunar reality has its own form of eternity, called effective eter-nity. This is demonstrated by the reproduction by which every living being seeks eternity not as a single individual, but as a species. Aristotle writes (De anima): "*The operation that for living beings is most natural of all (for those living beings that are perfectly developed and have no defects and no spontaneous generation) is to produce another being equal to itself: an animal an animal, a plant a plant, in order to participate, as far as is possible, in the eternal and the divine; for it is to that that they all aspire and that is the end for which they accomplish all that they by nature accomplish … Since, therefore, the living cannot participate in the eternal and the divine continuously, for the reason that none of the corruptible beings can remain identical and numerically one, then each participates in it to the extent that it is possible for him to participate, the one more, the other less, and there remains not him, but another similar to him, not one in number but one in kind".*

The sky is attributed a "numerical eternity". In fact, thanks to astronomical knowledge, according to Aristotle, all the motions of the celestial bodies can be placed on rotating spheres internal to that of the fixed stars and these spheres can be numbered.

The sky with its eternal cyclicity determines the alternation of seasons in the sublunar one, giving rise to the seasonal cyclicity in the effective eternity. The numerical eternity determines the effective eternity, but the vice versa is never possible.

The time seems destined to the "not to be" because as past it has been, but it is not anymore, while as future it will be, but it is not yet. Time, however, is not separable from movement, that is from becoming, it cannot exist if change does not exist.[2] It too is continuous, but in the succession (before and after) differently from space which is continuous in the extension.

In particular he says in the Physics: "…*when, in fact, we change nothing within our souls or feel that we change nothing, it seems to us that time has not passed at all*".

This shifts the philosopher's attention to the time–movement relationship, which gives the concept of time a greater concreteness.

Movement is in time which, in turn, does not exist except in movement. Time is "*the number of the movement according to the before and the after*".

He further adds: "*When … we think of the ends as different from the middle and the soul suggests that there are two instants, the before, that is, and the after, then we say that there is between these two instants a time, since time seems to be what is determined by the instant: and this remains as a foundation*".

All this highlights the number, that is, counting. Here he calls into question the entity in charge of this function: the consciousness, or soul, which

---

[2] Without wanting to sin of anachronism, the same type of problem is found in the condition of a system in thermodynamic equilibrium. When the properties of such a system are uniform and constant, it does not evolve and what about time? We will discuss this again in the part where we discuss the arrow of time.

is aware of this numerical succession of before and after in relation to individual life. In fact, he states: "*One could, however, doubt whether or not time exists without the existence of the soul. In fact, if the existence of the numerant is not admitted, the existence of the numerable is also impossible, so that, obviously, not even the number will be there. Number, in fact, is either what has been numbered or the numerable. But if it is true that in the nature of things only the soul or the intellect which is in the soul has the capacity to number, then the existence of time is impossible without the existence of the soul...*".

So, if there were no one (the soul) who is able to count there would only be the movement of bodies.

We take advantage here, not being able to do so in more detail for obvious reasons of space, to note how these same concepts were taken up by Saint Augustine (354–430 A.D.), albeit with a different perspective. The Saint says (XI book of the Confessions): "*I know what time is, but when they ask me, I cannot explain it*". Unlike Aristotle, he believes that time begins with Creation accomplished by a God outside of time. Time begins, therefore, at the act of creation. As we have already said, it tends not to be, but it cannot not to be because we perceive it and are able to measure its duration in intervals. Time is a continuous flowing from past to future and in some way, it exists thanks to the memory we have of the past and that memory recalls to the present. So also for present and future. Nothing clearer can be said than what Augustine himself says: "*One fact is now limpid and clear: neither future nor past exists. It is inaccurate to say that there are three times: past, present and future. Perhaps it would be correct to say that there are three times: present of the past, present of the present, present of the future. These three kinds of times exist in some way in the soul and I do not see them elsewhere: the present of the past is memory, the present of the present is vision, the present of the future is expectation*".

The reality of time can only be grasped from the interiority of the soul. Without the soul there is no time.

More than six centuries have passed between Aristotle and Augustine, although both have highlighted the great problematic nature of the concept of time. The former linked it to the concept of movement and both made use of the reference of the soul, that is, of a conscious entity capable of perceiving its passage. Aristotle recognizes that time comes from a cyclical reality, while Augustine's time could be said to be linear in that it proceeds from a beginning, creation, and goes towards an end, in accordance with Christian thought.

Another concept underwent an evolution and that was opinion. For Christians, the sensible world becomes a divine creation and, as such, reveals the manifestation of God in nature. Opinion is, therefore, a way of knowing God through knowledge of the phenomenological world. Already for Plato it was a first step towards the acquisition of truth. For Aristotle it had become a premise for scientific knowledge. Now it is an indispensable motion towards the knowledge of God.

2.3   The New Experimental Science

The combination of the conception of nature as the revelation of God's immanence and the revaluation of opinion as knowledge of the objective properties of sensible reality gave great impetus to the birth of experimental science. It will be entrusted with the task of verifying the adherence of reality to opinion. We begin this paragraph in a way that is perhaps unexpected for some in a text like this one. But we cannot fail to mention Leonardo da Vinci, if for no other reason than his great curiosity about nature as a whole and his capacity for observation and experimentation.

Leonardo (1452–1519) in his early childhood in Vinci was strongly influenced by his uncle Francesco. He introduced him to the careful observation of nature, following Aristotle's precepts on the concrete teaching of what exists in reality and to a deep love for all "Creation". This sort of imprinting accompanied Leonardo throughout his life, whether in his paintings, his intense and accurate study of the human body, the study of water and its flow, or various other physical phenomena to which he devoted himself in later life. Leonardo approached science with an essentially practical approach even though he anticipated ideas that were later supported by thinkers who, unlike him,[3] developed them organically. In fact it is said that: *"... all this will be done not by a scholar but by an artist. The conclusions reached by Leonardo during his incursions into the field of natural philosophy will therefore always take on a mediocre form from the point of view of the orthodox scientific treatise. However, this defect will be balanced by a unique outcome in the history of science: biological theories will be expressed for the first time not in verbal but in visual language and, above all, will become part of one of the strongest aesthetic visions of the world that man has ever known"* [3]. But we will say more about this.

We can, however, underline some fundamental aspects of the method of investigation applied by Leonardo which would later become typical of the scientific method. As well as documenting himself on the sources (today we would say bibliographical analysis of the subject to be tackled) he dedicated

---

[3] The collection of Leonardo's notebooks is perhaps the best we have to learn about his studies and philosophical ideas. It has already been partially digitized by the Victoria & Albert Museum in London.

himself to systematic observation and combined it with careful experimentation. Systematicity, which also implies repetition, allows a detailed acquisition of the characteristics of the observed and experimentation, when possible implies the need to replicate the phenomenon and is a confirmation or not of its correct understanding.

All this was followed by a series of measurements. The measurement of space and motion accompanied Leonardo throughout his life, whether it was dedicated to the creation of paintings, the study of the motion of water, machines, anatomy or urban planning.

Finally, there was the search for the use of mathematics as a tool for representing the order and proportions of the universe. Mathematics, therefore, as a tool to better understand and make rigorous physical observations. To this was added, and was an essential part of it, a large set of exceptional visualizations.

It is somewhat reminiscent of a modern scientist who uses high-speed, high-resolution optical visualization systems for his studies, fixing images from which to draw insights and the subsequent construction and validation of mathematical models for the interpretation of phenomena.

But nothing more because, as some have said, Leonardo's science was a science of quality and of forms with their movements and changes. Mathematics described the continuous transformation of geometrical forms that obeyed the rigorous and coherent laws according to which nature was ceaselessly shaping them. There is a dynamic conception of the world in which everything is transformed, which will consequently influence the ideas that Leonardo will manifest towards time.

Having said this about the novelty and importance of the method of investigation that emerges from the Master's work, let us return more specifically to the subject at hand. Leonardo took the opposite position to Aristotle on the motion of bodies. Aristotle, in fact, believed that all motion originated directly from the "Prime Mover", whereas for Leonardo matter has an innate tendency to move unless it is stopped. He says: "*Motion is the cause of all life*". He thus anticipated the concept of inertia which Galileo established about half a century later. Space, in turn, stands in relation to art as the representation of dimension and relief and from this comes the dispute between painting and sculpture as the best form of such representation.

Time, on the other hand, has much broader and universal implications. For art, it represents the time when it takes place and when it is enjoyed, but also the "enemy" to be challenged for the realization of a beauty that is saved from death. It is not only an element of comparison for every human activity of which art is a manifestation, but also for what happens in nature.

Nature is the site of a continuous struggle between the tendency of "forms" to aggregate and the force of "consumer" time that dissolves them. As Leonardo says in the Arundel codex: "*Oh time, consumer of things, in you turning them, you give to the tracts new and various lives. Oh time, swift predator of created things, how many kings, how many peoples hast thou undone*".

We will return to this idea of consumer time again in a much more recent light.

We have seen how Leonardo thought of a living nature in continuous change and movement. Rejecting an a priori acceptance of the dictates of the classics, he used an empirical approach to verify through observation their effective adherence to reality and he used to say: "*wisdom is the child of experience*". His eye was that of the artist, but it also became the investigating eye of the scientist. Art and science are in him a whole and can hardly be fully understood by separating one from the other.

This vision of the dynamism of reality, of the infinite variety of living forms in continuous transformation and self-organization (think of the growth and evolution of various phenomena, such as the flow of water in a river), of their mutual interrelationships (for example, those between the organs of a human body) led to the definition of Leonardo as a theorist of complexity, a science that has developed only in recent times.

He was also convinced that nature was far superior to the machines designed by man and that the correct attitude towards it was to love it and learn from it and not to dominate it.

Before leaving this character we like, in the next reading, to dwell on a simple experiment, that anyone can easily replicate, in which Vinci's scientist shows an exceptional ability to observe phenomena together with a great skill, moreover usual in him, in describing them combining the verbal comment with the effectiveness of his extraordinary drawings.

Leonardo and the Candle [4]

In a bifolium of the Atlantic Codex we find Leonardo's observations on the interactions between the flames of several candles and on what we would call the fluid dynamics of the combustion of a tallow candle flame. Let us dwell on this last aspect, described in Fig. 2.2, on the left with a reworking of Leonardo's drawing and on the right with a candle flame around which lines are drawn that schematize the movement of air.

Leonardo begins by explaining the principles that underlie the phenomenon with an essentially Aristotelian approach. He explains that smoke rises upwards because it is hot, being produced by combustion. As it rises it cools and falls back down.

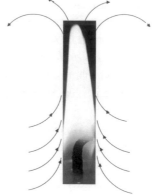

Fig. 2.2 Illustration of Leonardo da Vinci's candle experience

Let us leave the word to Leonardo making some remarks now and then. *"The blue smoke is a transit of the material nourishment of the flame of such a candle* [the tallow of the candle burning forms a small globule at the base of the flame which Leonardo indicates as blue.] *The alb* [white] *smoke surrounding the remainder of the flame is spiritual transit* [spiritual is in contrast to material. This is due to the burning of the tallow and produces the smoke. Having consumed the tallow in smoke, a white area is formed around the blue globule.] *The igneous vapour that is infused in the moist smoke... clouds the end of the narrow flame which* [high end of the flame] *for this reason becomes reddish and to the last of dark redness...* [then he pays attention to the movements of the air around the flame] *The air that, from outside, continuously strikes the flame and pushes it back, condenses and this condensation makes the flame brighter and more resplendent, and the condensed smoke breathes through the upper part of the flame and does not succeed in any other place because downwards it finds the matter that generates it* [it does not go down because it is blocked by the matter of the candle] *and from the sides it finds the air that strikes it, and above it finds the dilation of the air that cleaves and by this dilation the smoke has its outcome ... Between the flame and the candle a revolution of air is generated which, after it has affected the base of the flame, turns downwards and, thus heated, strikes the front* [the base] *of the candle and resolves it and prepares it for the nourishment of the aforesaid flame"*.

At first, he gives a chromatic description of the various areas of the flame with the careful observation of the painter. In fact the colouring depends on the various temperatures of the flame. Leonardo does not know this, nor does he care, but this is how he looks at nature, accustomed as he is to representing it. He then goes on to describe the movement of the surrounding air, which strikes the flame both at the base, participating in the combustion, and contributing to its shape by striking it. Finally, the air, heated by the candle, first rises and then, as it cools, descends with a whirling movement to return to the wick and once again fuel combustion. These are typical observations of anyone who today studies, mutatis mutandis, the fluid dynamics of flames.

After having only briefly outlined the thought of Leonardo da Vinci, we do the same for another great scientist, Galileo Galilei, again limiting ourselves to a few elements closely related to the topic we are talking about.

Galileo (1564–1642) is unanimously indicated as the founder of the experimental method. As we have seen, the method consists in verifying scientific hypotheses by means of experimental procedures that may or may not confirm their validity. Theorized by Francis Bacon (1561–1626), it was systematized by Galileo and further perfected in the following centuries.

Galileo makes an initial distinction between the true Aristotle and the false Aristotelian school. To the former he attributed a real interest in the sensible world, of which he was a careful observer, to the latter he attributed that it had not added anything in terms of new experiences, but simply commented on the results of the master on the basis of a principle of authority.

The experience must be carried out directly and repeated several times, since an occasional and/or superficial verification is not sufficient. It must be followed by reasoning, which is sometimes sufficient in itself to solve a problem. Only in this way can we increase our knowledge of sensible reality, while uncritically relying on the principle of authority leads us to construct a false reality. Here we must not forget that Galileo could avail himself of knowledge and instruments unknown to Aristotle.

Mathematics is the language with which nature speaks to us, and the mathematical theories that interpret experimental observations describe reality in its substantial and enduring aspects. He too sees mathematics not as an end in itself but as a means of understanding nature. So much so that it is measurable in number weight and size, or rather such are the bodies that constitute it. This is the characteristic of their primary qualities, that is, their form, their relationship with other bodies, the space they occupy and the state they are in: still or in motion. Colour, smell, taste and all the other characteristics that depend on our way of perceiving them are part of the secondary qualities, as they do not contribute anything to the real properties of the bodies themselves. Among the measurable quantities, weight induced Galileo to affirm that the only natural motion was the downward motion of bodies and that the characteristics of heaviness or lightness depended on the medium in which bodies are immersed, as deduced from Archimedes' hydrostatics, and consequently could not be intrinsic qualities.

The movement was in open contrast with the Aristotelian view. Let us recall that in the meantime Copernicus (1473–1543) had introduced the heliocentric theory, already discussed by Aristarchus of Samos in the third century B.C., in sharp contrast with the Aristotelian–Ptolemaic geocentric model held until then (the book "De revolutionibus orbium coelestium", in which he argues for it, was published the same year of his death). Both Aristotle's and Ptolemy's cosmological models, which placed the earth, and hence man, at the centre of the universe, were replaced by one that instead placed the sun at its centre. In addition to revolutionizing astronomy, he also upset the anthropocentric philosophical and theological conception. Galileo, convinced that science should be independent of faith, supported Copernicus' vision and, as we know, paid the price for such a "misdeed". The conclusions he reached through observation and reasoning were of enormous significance. In his youth, during his stay in Pisa, the focus of his attention was pendulums. In particular, he understood that the weight of a pendulum during an oscillation, in the absence of friction, would return to the same height from which it had fallen. An intuition that today we would translate by saying that, in the absence of causes of dissipation of its initial energy (depending on the height from which it starts, potential energy), this is conserved and the pendulum oscillates indefinitely from one side to the other. He also understood, accurately applying his experimental method, that the speed with which a body rolling on an inclined plane reaches the ground depends only on the height from which it is dropped and is not influenced by the inclination of the plane. In the same way two bodies of different weights dropped in the same instant, in absence of friction, take the same time to reach the ground. This means that, in the absence of forces opposing the motion (air friction forces), the weight (and the shape) of the body has no influence on the time it takes in free fall to cover a certain distance (from the point where it is dropped to the ground), so time and space are connected by a relation that does not include the characteristics of the body.[4] Contrary to

---

[4] Whoever remembers a bit of physics knows that the space, s, covered by a body in free fall, with zero initial velocity (we do not throw it towards the ground, but we simply let it fall), and left at the instant $t = 0$ is given by (s is the distance from the ground):

$$s = -\frac{1}{2}gt^2 + h$$

If we let it go from a height h relative to the ground, which is at elevation zero, the space it will have to travel will be just h and the time taken in the entire path will be:

$$t = \sqrt{\frac{2h}{g}}$$

It is clear that in this relationship the characteristics of the body do not appear, but only the height and the acceleration of gravity. If there were air friction, the shape of the body would come into play. How many times, in fact, do we hear talk of an aerodynamic shape, with a low coefficient of friction, for automobiles and airplanes. Taking advantage again of the reader's knowledge, we also remember that under the same conditions the velocity is $v = gt$, that is, proportional to time. Galileo established precisely that the speed of fall increases proportionally to time and that the corresponding acceleration is the same for all bodies.

Aristotle, who believed that the natural state of bodies was stillness and that motion was ultimately determined by the Immobile Motor, Galileo thought that bodies can be in uniform rectilinear motion even without the action of forces. Rather, these forces modify their state of stillness and motion. The natural state is not stillness, but motion. Finally, he enunciates the so-called principle of relativity, according to which the laws of mechanics remain the same on any system that moves with uniform rectilinear motion, called inertial system. If we are on such a system, so we are moving at constant speed on a rectilinear trajectory, we are not able to say if we are in motion or stopped. Unless we can observe external reference points; but even then we can only say that we are in motion relative to them. In an airplane you have surely observed the difference when the stewardess pours your coffee, during a calm flight (the plane is at cruising speed and everything is quiet) and when there is turbulence. In the latter case, you risk getting a bad stain; in the former you do not. Correct, but why? If everything is calm you are on a reference that moves in a straight and uniform way and the trajectory of the coffee from the thermos to the cup is the same that it would follow in a bar on the ground. If there is turbulence all this is no longer true because there are forces that modify the laws of motion from the pot to the cup. Galileo tells us the same things referring to a ship. We report here the passage from the "Dialogue on the two greatest systems of the world" (1632): "*Reserve yourselves with some friends in the largest room that is under the cover of a large ship, and here try to have flies, butterflies and similar small flying animals; there should also be a large vase of water, and inside it some small fish; hang some buckets high up, which drop by drop will pour water into another vase with a narrow mouth, which is placed below; and while the ship is standing still, observe diligently how those little flying creatures go with equal speed towards all parts of the room; the fish will be seen to go noticing them indifferently in all directions; the falling drops will all enter the vessel below; and you, throwing something at your friend, should not throw it more forcefully towards that side than towards this one, when the distances are equal: and jumping, as they say, on foot, you will pass equal distances towards all parts. Let the ship move with whatever speed you wish; so that (even if the motion is uniform and does not fluctuate here and there) you will not recognize the slightest change in all the effects mentioned, nor will you be able to tell from any of them whether the ship is moving or standing still*".

On the basis of what we have said if we are in motion on an inertial system, we are not able to say if it is moving or stationary with respect to the earth, unless we can see houses signs or trees. In this case we can say if we are at rest or in motion, but only with respect to the earth. By saying that a train is going 120 km/h strictly speaking we are saying something at least inaccurate, we should say that its speed is 120 km/h with respect to the ground. If we overtake a lorry travelling along a road parallel to the railway at 80 km/h our speed relative to the lorry is 40 km/h. All this constitutes our established knowledge and we obviously agree with Galileo. We probably agree with him about time, because we too are used to considering it an absolute physical entity, measurable and associated to space by movement. In fact, both Galileo and, later, Newton spoke of an absolute time, objective precisely because it can be measured with instruments, and a subjective time, not measurable, but internal. Newton (1642–1727) in fact supplies at the beginning of his Mathematical Principles of Natural Philosophy some definitions that we report for their clarity. These concepts are still the bases of the classic mechanics.

"… I do not define, instead, time, space, place, and motion, because they are well known to all. It must be observed, however, that these quantities are commonly conceived only in relation to sensible things. Hence arise the various prejudices, for the elimination of which it is necessary to distinguish the same things in absolute and relative, true and apparent, mathematical and vulgar". So, Newton divides these quantities into two types: the absolute one, which he indicates as true and mathematical, and the relative one, which is apparent and vulgar. And he goes on with a sort of list:

"I. Absolute, true, mathematical time, in itself and by its nature without relation to anything external, flows uniformly, …; relative, apparent and vulgar time is a (accurate or approximate) sensible and external measure of duration by means of motion, which is commonly employed in place of true time: such are the hour, the day, the month, the year.

II. Absolute space, by its nature without relation to anything external, always remains equal and immobile; relative space is a mobile dimension or measure of absolute space, which our senses define in relation to its position with respect to bodies, … Absolute space and relative space are identical in size and species, but do not always remain identical in number. For if the Earth, for example, moves, the space of our air, which relatively to the Earth always remains identical, will now be a part of the absolute space through which the air passes, now another part of it; and so it will change absolutely in perpetuity".

Absolute time and space exist in themselves as such independently of the observer, while relative time and space cease to exist in the absence of the observer and, therefore, depend on the observer. In addition to the relativity of motion emerges an idea of time that today we would divide into physical and psychological.

"III. Place is the part of space occupied by the body, and in relation to space it may be absolute or relative.

IV. Absolute motion is the translation of the body from an absolute place into an absolute place, the relative from a relative place into a relative place. We define, in fact, all places by the distances and positions of things in relation to some body, which we assume to be motionless; and then with reference to the aforesaid places we evaluate all motions, inasmuch as we consider the bodies to be transferred from those same places to others. Thus, instead of absolute places and motions we use relative ones; nor is this inconvenient in human affairs, but in philosophy it is necessary to abstract from the senses".

The way we see things, even after studying classical mechanics, is basically just that. Even those who have studied a little physics have the habit of making graphs in which time appears as an independent variable. We report as a function of this, which would be Galileo's and Newton's absolute time, the speed and space travelled by a bullet (to cite a common example in physics courses), the growth of a plant and an infinity of other things. At the same time, we feel time flow differently according to events. We will see, however, that these certainties about absolute time will be strongly shaken when we will talk about the propagation of light and phenomena that occur at speeds comparable to that of light itself.

References

1. Horowitz, A. (2017). *How your dog thinks—All the secrets of man's best friend.* Mondadori, Oscar Essays (in Italian). In English: (2009). *Inside of a dog: What dogs see, smell, and know.* Scribner.
2. Rindone, E. (2003). This text and the two that complete the path—"The Problem of Time and History in the Bible" and "The Problem of Time and History in Medieval Philosophy", resume the papers given at the August 2003 Norcia Seminar on the theme "The Human Adventure between Time and History" Sixth Philosophical Week for ... non-philosophers (in Italian).
3. Laurenza, D. (1999). Leonardo, Science transfigured into art. *Le Scienze,* (Scientific American) Year II, n. 9, Milan, p. 33.
4. Leonardo and the flame of the candle, by Paolo Galluzzi - Museo Galileo, EDUCAZIONE SCIENTIFICAB-10-FSE-2010-4. Accessible on: ©INDIRE 2014, http://formazionedocentipon.indire.it. Also viewable on YouTube (in Italian). In English: https://www.youtube.com/watch?v=tn03h1AcspQ. Accessed November 2021.

3

Charges, Magnets and Light

3.1 Moving Charges and Magnetism

It is not unusual for scientific discoveries to occur along sometimes random paths or at the hands of men who by birth seem destined for something else. It is the story of Michael Faraday (1791–1867) born in Southwark, a village near London. He was the son of a blacksmith and at thirteen was sent to work for a bookseller as an assistant bookbinder. Only, instead of just binding books, he also read them. Although he had a rudimentary education, he was interested in electricity and chemistry. He also took advantage of the bookbinding workshop to perform some simple scientific experiments. He intrigued a customer who gave him tickets to the Royal Institution in London to attend lectures by Humphry Davy (1778–1829), a professor of chemistry. After some time, Faraday was hired by Davy as a laboratory assistant, at first put to the test with humble tasks. Later, he had the unique opportunity to accompany his mentor on a long trip to the greatest European cultural centres, meeting famous physicists such as Ampere and Volta, who greatly appreciated his qualities. On his return, he resumed his work in chemistry for the Royal Institution, with which he maintained a constant link throughout his life. Faraday made substantial contributions to chemistry, for example isolating magnesium and finding the two laws that bear his name in the field of electrochemistry. In fact, he would have become famous if only for his various contributions in these areas, although today he is better known for his discoveries in electromagnetism. This is the aspect that we will examine, because it is closely related to what we will say next.

W. Grassi, *The Challenges of Time*,
https://doi.org/10.1007/978-3-030-94372-1_3

Before going on, let us briefly take stock of the knowledge that already existed and of that which was being formed in Faraday's time, with regard to the subject we are dealing with. It has been known for a long time that two electric charges repel or attract along the conjunction of their centres, depending on whether they are of the same or opposite sign. The direction was the same according to which two masses attract. It was also known that magnets had a positive and negative pole and that poles of opposite sign attracted each other while those of the same sign repelled each other. Besides these aspects typically inherent to electric and magnetic phenomena, they tried to measure the value of the speed of propagation of light. As we will see, this too is part of electromagnetism. After an experiment attempted by Galileo himself who intended to measure it by means of two lanterns placed at a distance of 1 km, the first measurements were made by Ole Roemer (1644–1710), a Danish astronomer, observing the motion of Jupiter's moons, as assistant to Giovanni Cassini (1625–1712) in Paris. Cassini had already thought, with similar observations, that the speed of light had a finite value. Roemer estimated it at about 214.000 km/sec, because of the inaccuracy of the data at his disposal. Later, Hyppolite Fizeau (1819–1896) and Leon Focault (1819–1868), who performed experiments starting in 1849, established that this value was approximately 300.000 km/sec.

In 1820, the Danish physicist Hans Christian Oersted (1777–1851) discovered that when he brought the needle of a compass close to a wire with a strong current flowing through it, the needle moved in a plane perpendicular to the wire as in Fig. 3.1.

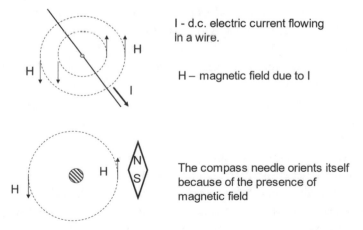

I - d.c. electric current flowing in a wire.

H – magnetic field due to I

The compass needle orients itself because of the presence of magnetic field

Fig. 3.1 Orientation of a magnetic needle in the presence of an electric current

Unlike the direction of attraction (or repulsion) we have discussed, the current seemed to exert an action orthogonal to its movement. A decade later, Faraday, convinced that there must be reciprocity between the effects of currents and magnetic fields, moved a magnet close to a coil of electric wire. He observed the passage of a current through it. The current was generated only when the magnet moved; i.e., it varied the magnetic flux (which is a magnetic field multiplied by an area) affecting the coil. If in Oersted's experiment moving charges produced a force on a magnet, now the movement of a magnet caused a force on the electric charges of the wire, setting them in motion and generating a current. It was sufficient that there was a relative motion between magnets and coils. So, he discovered the phenomenon of magnetic induction. If you use a bicycle with lights, which is something that is increasingly rare, and you power them with a dynamo, you have an easy example of the application of this principle. A type of dynamo for bicycle is shown, very schematically, in Fig. 3.2: a cylinder of magnetic material (e.g. iron) on which are wound some coils of electric wire rotates inside a field produced by a permanent magnet. From the coils, which are connected to an electrical collector with devices called brushes, the current circulating in the coils is collected and sent to the lights. On the same principle, although in this case the current is alternating and the magnetic field rotates, are based the machines that convert fuel energy into electricity in power plants (alternators, also used in cars), as well as those called transformers that have an infinite number of uses.

There remained the problem of explaining how these actions occurred at a distance. We are used to applying forces to objects, but to do this, we have to

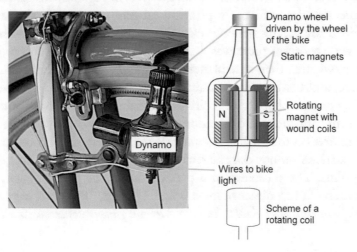

Fig. 3.2 Dynamo of a bicycle

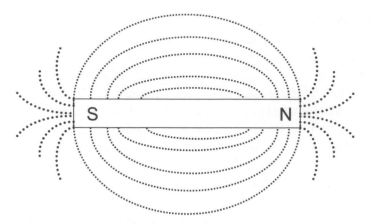

Fig. 3.3 Diagram of the arrangement of iron filings in the presence of a magnet

touch them. If we want to move a parcel, we pick it up with our hands and put it somewhere else: we need physical contact. But Faraday had observed how iron filings were placed on a sheet in the presence of a magnet. It was exactly the same kind of experiment that is shown in high school physics classes. He had noticed that it was distributed along lines like those shown in Fig. 3.3.

It seemed that, point by point, there were forces that arranged the fragments of filings according to a certain geometry. Faraday called them lines of force. Due to the presence of the magnet, force fields were formed in space, which, in this case, could be visualized with the iron filings. Also in the case of Oersted's experiment, the current flowing through the wire generated a force field (the circumferences concentric to the wire in Fig. 3.2) which caused the displacement of the needle in a position tangent to them. In the same way, we explained the actions between electric charges and between masses. Until then, as we have seen, there was a separation between the concept of space and bodies, whether they were planets or atoms. Space extended continuously everywhere, and bodies were contained in it. Faraday argued that we should forget the concept of space as a single continuous entity that contains and isolates atom from atom and abandoned the idea that there was space on one side and matter on the other. He contradicted, therefore, the atomic theory.[1] Material particles do not exist as entities of a certain size and actions reside around points. The substance of a particle is not the particle itself, but the action it exerts from a dimensionless point. He said, "*Why must we assume the existence of something we do not know, which we cannot conceive, and for which*

---

[1] According to this, atoms are discrete objects separated by an empty space.

there is no philosophical necessity?". In this way, he proposed his continuum theory with a conceptual leap from continuous space to continuous matter that fills all space.[2] Gravitation itself is a "quality" of matter, and it is the gravitational force that constitutes matter. With all this, actions at a distance are still incomprehensible, we only know something more about the properties of space (matter) that separates bodies. Keep in mind that Newton himself, and Faraday was a Newtonian, had argued that his law of gravitation needed a material medium between bodies. Time becomes a fundamental element of his theory. He contrasted, in fact, the idea that actions at a distance occur instantaneously and he did so, for example, by referring to the propagation of light, saying "*The propagation of light, and therefore probably of all radiating actions, occupies time; and, in order that a vibration of the line of force may explain radiating phenomena, it is necessary that such a vibration should also occupy time*". Light propagates in time like all radiant phenomena, and, as we have seen, this was a concept that was becoming established thanks to the experimental verifications of various scientists. Time is, in general, an essential element in the evolution of interactions and, therefore, in the action of forces. Faraday tried until the end, without success, to perform an experiment to show that the lines of force propagate in a finite time, in order to give an experimental evidence to his field theory.

The concept of field, to which we are accustomed and which we talk about every day (think of how much we talk about electromagnetic fields, from smartphones to television to microwave ovens), is fundamental in physics. Basically, we can say that the presence of a magnet, an electric charge or a mass, changes the properties of space by producing force fields there. Faraday, because of his lack of mathematical knowledge, was unable to translate his discoveries into formulas, which caused him the hostility of many, even though they did not know how to explain it.

---

[2] Faraday, therefore, maintains that matter consists of atoms that are points (in the mathematical sense of the word) with a sort of atmosphere of force around them. He partly adheres to the thought of R. J. Boškovič (Dubrovnik, 1711–Milan, 1787), a Jesuit who dealt among other things with physics and astronomy and who, however, unlike Faraday, distinguished between force and matter according to Newton's precepts.

3.2   The Electromagnetic Field and the Lorentz Hypothesis

It was James Clerk Maxwell (1831–1879), a Scottish mathematician and physicist, who built the theory that translated Faraday's thought into mathematical terms. Unlike Faraday, he came from a noble family and at the age of sixteen became a student at the University of Edinburgh. In addition to the theory of the electromagnetic field, his studies also led him to formulate the kinetic theory of gases. In 1860, he became a professor at King's College in London. It was a very prolific period in which he dealt with thermodynamics, the kinetic theory of gases and, indeed, the theory of the electromagnetic field. In this same period, he came into contact with Faraday. In the meantime, the British had begun to take an interest in signal transmission. Already in 1844, thanks to the invention of the alphabet of the same name by Samuel Morse (1791–1872), the first telegraph line between Washington and Baltimore had been activated in the USA. This system of transmission spread rapidly through the nations, giving rise to increasingly complex networks. At the same time, they began to connect countries separated by the sea. The first attempt was made in 1845 with a submarine cable of about 2 km inside the bay of Portsmouth and, five years later, connecting Dover to Calais (after a few days the cable was accidentally broken by a fisherman). Both these projects were carried out by British companies and began the development of submarine cable networks until, around 1865, Ireland and Newfoundland were connected. Thanks to the interest in applications and the studies of various scientists, including Faraday himself, the number of proofs that there could be a link between electrical and optical phenomena was growing. Maxwell demonstrated in the treatise "A Dynamical Theory of the Electromagnetic Field" that the electromagnetic field propagates as waves at the speed of light and that, therefore, light is an electromagnetic phenomenon. Contrary to Faraday's view, according to which there was no necessity for the existence of the ether, he remained anchored to the existence, not measurable, of this medium permeating empty space.

Waves and the Ether

The concept of wave is familiar to us at least since the first time we went to the sea. Moreover, if we have thrown a stone into the still water, we have seen that from the point where the stone falls waves are formed that move away from it to propagate on the surface. There are also waves that we do not see, but that we hear: sound waves. In the case of water, waves correspond to a vertical movement of the surface that rises (crest of the wave) and lowers (belly

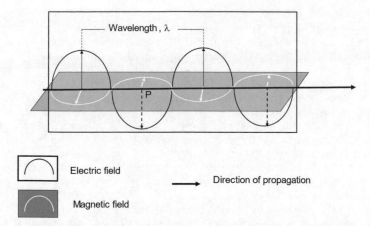

Electric field

Magnetic field

Direction of propagation

Fig. 3.4 Propagation of electromagnetic waves passing through a generic point *P* at a generic time *t*. The electric field and the magnetic field vibrate on axes that are orthogonal to each other and propagate in a direction perpendicular to them

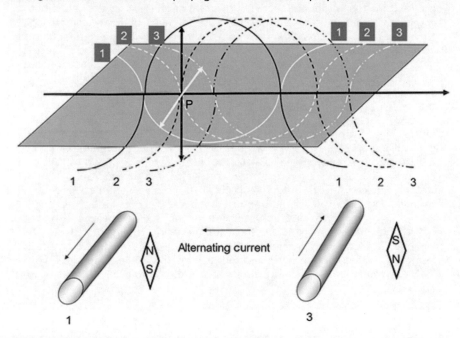

Alternating current

Fig. 3.5 Passage of an electromagnetic wave in a point *P*. The directions of the current flowing through the wire are indicated with 1 and 3 referring to two successive instants. At instant 2, the current is null

of the wave) advancing. The direction of propagation is perpendicular to the wave, that is, to the rising and falling of the surface. The distance between two successive crests (or bellies) is called the wavelength. These are called transverse waves and behave similarly to the electromagnetic field. The ones we hear are sound or acoustic waves and are vibrations of air pressure. Think of a piston moving forward in a cylinder. As it moves, it moves the air in its own direction of motion. The gas particles closer to the piston press on the others, becoming denser in some areas (ridges) and thinner in others (bellies). They are called longitudinal waves because the direction of propagation, which occurs in the direction of motion of the piston, coincides with that of vibration, according to which the crests and the bellies are formed. The speed of propagation depends on the nature and conditions of the medium in which they move. For example, sound waves in air propagate at 331 m/sec (1191.6 km/h) at a temperature of 0 °C and 343.8 m/sec (1237.7 km/h) at 20 °C, in water at about 1500 km/h and in granite at 6200 km/h. What do these two types of waves have in common? Both need a medium, water or air, in which to propagate. In space, noises cannot be heard. The same electromagnetic waves are affected by the conditions of the medium, but they can also propagate in vacuum and they do it reaching the maximum speed. For this reason, there arose the necessity to conceive the ether, which was connected with the need to identify, in analogy to the types of waves already known, a medium within which electromagnetic waves could act. An example of electromagnetic wave, in which the electric and magnetic fields are concatenated, in the sense that they are deeply related one to the other, is shown in Fig. 3.4. The electric field oscillates on a plane, the magnetic one on an orthogonal one and the direction of propagation is that, in the case of light waves, of the light ray.

In the light of what we have said, what does the magnetic needle do in the presence of a sinusoidal alternating current, like the one we have in our homes? Let us fix our ideas on a point P on the axis of propagation. The electric wire is placed perpendicular to it, and the whole is drawn in Fig. 3.5. The numbers 1, 2 and 3 indicate three successive instants. At instant 1, the current is directed in the outgoing direction from the sheet that is towards the reader. At 2, its value is null and at 3 has the same value as 1, but it is directed in the opposite direction, away from the reader. The values of the electric field that produces it are those of the white curves, as before. The magnetic field behaves as usual so that at instant 1 the magnetic needle is oriented in one way and at 3 in the opposite way. This is what happens if with a compass you approach a conductor crossed by alternating current, like those of power lines.

In any case, his unifying theory of electric and magnetic phenomena, experimentally verified by Heinrich Rudolf Hertz (1857–1894) in 1887, remains a milestone for physics. Einstein himself said [1]: "*It was then that the great upheaval took place, to which the names of Faraday, Maxwell and Hertz will always be linked; but it was Maxwell who had the lion's share of this revolution. He showed that what was then known about light and electromagnetic phenomena is represented by his well-known double system of partial differential equations, in which the electric field and the magnetic field intervene as dependent variables. Maxwell actually tried to give a basis to these equations or to justify them by means of the ideas of mechanics. … Before Maxwell, physical reality was imagined (as representing the phenomena of nature), as material points whose modifications consist only in movements, regulated by partial differential equations. After Maxwell physical reality was conceived as represented by continuous fields, not mechanically explainable, regulated by partial differential equations. This change in the conception of reality is the deepest and most fruitful change that physics has undergone since Newton; …".*

What does this theory imply for the concepts of space and time? Maxwell was deeply convinced that, in spite of the close correlation that exists between the human intellect and the laws of nature (we obviously use our intellectual capacities to create theories and interpret them), we must operate by performing an objective control in the formulation of the concepts. He did not therefore refer to time and space as absolutes in the same way that, for example, Newton had done. His equations showed that there is a continuous field, at the basis of physical phenomena, whose distribution can be described at any instant and at any point in space. The finite speed with which the field changes highlights the empirical role of space and time, based on experience. The same equations have a characteristic fundamentally different from those of mechanics. The propagation of the electromagnetic field does not obey the Galilean relativity. Remember, in fact, that an observer in uniform rectilinear motion sees a glass fill in the same way as when it is stationary (always with respect to the ground). This means that the equations that describe the phenomenon remain the same changing inertial reference or, as they say, they are invariant with respect to this change. Maxwell's equations do not enjoy this property and are, in this sense, asymmetrical with respect to those of mechanics. Two observers in motion with a given relative velocity will see differently an electromagnetic phenomenon. This aspect will be harbinger of great innovations in physics. In the meantime, it led to exclude the existence of the ether, thing confirmed in time, but at first supported by Albert Abraham Michelson (1852–1931), of Prussian origin, and Edward Morley (1838–1923), American, with some experiments performed since 1887. The hypotheses they made and the real meaning of their results were debated for many years by the scientific community. For his invention of the interferometer with which he conducted his experiments and for his research in optics, Michelson received the Nobel Prize for Physics in 1907. Let us outline the experiment rudimentarily, in a nutshell. They hypothesized the ether fixed relative to the sun. If the earth moves with respect to the ether, it produces in it a movement in the opposite direction, which was called "ether wind".

To explain further let us take an example with automobiles. Many have heard about CX, that is the fluid dynamic resistance of these to the air. In order to reduce fuel consumption, we try to reduce it by making aerodynamic bodies and, in racing cars, we mount skirts and wings that improve the adherence to the ground. The studies are carried out in the so-called wind tunnels, large laboratories in which air is blown with large fans onto the stationary and suitably instrumented car. Instead of making the car move through the air at a certain speed, the air is moved with respect to the car at the same speed. Thus, the same relative velocity is maintained between the car and the air. A body moving in the ether would trigger in it a motion interpretable in an analogous way. Given the non-invariance of the electromagnetic field to Galilean transformations, the times employed by two light rays (moving, obviously, with the velocity of light) sent in two different directions will have to employ different times[3] in covering the same distance. They considered a direction parallel to the motion of the earth, in the same direction, and a perpendicular one. With mirrors placed at the same distance from where the signal was sent, they reflected the signal back to the starting point. They noticed that in both paths the time taken was the same.[4]

At this point, we need some math to better see what the experiment consists of, at least in a very schematic way, and what the implications of this result are.

The experimental apparatus is drawn in Fig. 3.6. A monochromatic light source, such as a laser, sends a beam to a semitransparent mirror (S) inclined at 45°, which divides it into two equal beams one in the same direction as the motion of the earth and the other perpendicular to it. The first one is reflected by a mirror, indicated with B in the figure and the second one by another one, indicated with A. The two rays return to it and together reach a receiver. Keep in mind that if the earth moves with respect to the ether with a velocity v, the ether moves with the same velocity, but in the opposite direction, $-v$, with respect to the earth. Let us examine separately the two rays.

Ray parallel to the motion (S–B and B–S paths).

In the path from S to B, moving, for hypothesis, the ether with respect to the earth with velocity $-v$, the ray moves with velocity $(c - v)$, while from

---

[3] The difference in times would be observed due to the formation of interference fringes at the point where the reflected rays returned. Hence the name interferometer given to the instrument. This had two arms of about 11 m and amplified the phenomenon by repeating the reflection eight times.

[4] No fringe interference occurred.

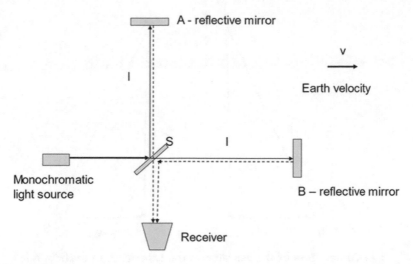

Fig. 3.6 Scheme of the equipment used by Michelson and Morley

B to S with velocity $(c + v)$. In the complete path from S to B and back to S, the time taken by the light ray is:

$$t_{S-B-S} = \frac{l}{c - v} + \frac{l}{c + v} = \frac{2cl}{c^2 - v^2} = \frac{2l}{c\left(1 - \frac{v^2}{c^2}\right)}$$

Ray perpendicular to the motion (S–A and A–S paths).

Now things get a bit more complicated, because the motion is orthogonal to the ray of light. The ray starts from S at a certain instant and reaches A when this has moved to a different position following a straight path but inclined with respect to the perpendicular to the motion. The same happens for the ray reflected by A, as shown in Fig. 3.7. In it we indicate the vectors (i.e. why they are in bold) of the velocity of light, $\mathbf{c}$, of the ether, $-\mathbf{v}$, and their resultant, $\mathbf{v_R}$, that is their vector sum. The space is now crossed with a velocity equal to $\mathbf{v_R}$ with value given by, being the triangle of the velocities

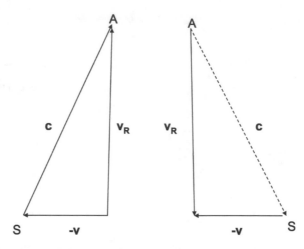

Fig. 3.7 Composition of the velocities in the experience of Michelson and Morley

rectangular:

$$v_R = \sqrt{c^2 - v^2}$$

It is the same on the outward and return journey, and the space covered is $2l$ in length. The time taken is therefore:

$$t_{S-A-S} = \frac{2l}{\sqrt{c^2 - v^2}} = \frac{2l}{c\sqrt{1 - \frac{v^2}{c^2}}}$$

Before going on, let us fix our ideas on some aspects. First of all, Maxwell's equations say that the speed of light in vacuum has the same value in all directions,[5] and for this reason, we have considered the same value both for the direct ray (towards the mirror B) and for the ray perpendicular to it. The speed of the earth around the sun during an orbit changes in direction (like when you were a child on the merry-go-round). For this reason, Michelson and Morley rotated the apparatus ninety degrees. Later experimenters, with more precise instruments, found the same results by rotating the apparatus variously.

---

[5] The ether cannot be distinguished from vacuum. The same happens in any homogeneous medium, even if the speed of light may have different values in media in which the relative permeabilities, electric and magnetic, are not unitary.

Let us look again at the relationships we found for travel times and see what conclusions we can draw from them.

$$t_{S-B-S} = \frac{2l}{c\left(1 - \frac{v^2}{c^2}\right)}$$

$$t_{S-A-S} = \frac{2l}{c\sqrt{1 - \frac{v^2}{c^2}}}$$

Without knowing how to read and write, as they say, we could immediately affirm that the earth is stationary with respect to the ether ($v = 0$) and that, for example, the hypothesis made of the ether integral with the sun is not true. Let us try for a moment to exclude this possibility also because by now the existence of the ether is no longer a necessary hypothesis to the explanation of the phenomena of electromagnetic propagation. Let us try instead a way that at the moment many people will consider at least brainy and twisted. Let us observe meanwhile that in the calculation of the velocities we used the Galilean principle of relativity valid in classical mechanics, as far as it was known then. Suppose also, for a moment, that the lengths of the arms of the interferometer are different l_B in the path from S to B and l_A in that from S to A, but not known both. We would then write:

$$t_{S-B-S} = \frac{2l_B}{c\left(1 - \frac{v^2}{c^2}\right)}$$

$$t_{S-A-S} = \frac{2l_A}{c\sqrt{1 - \frac{v^2}{c^2}}}$$

We know that we have measured the times in a system integral with the experimental apparatus (we will see later the meaning of this observation). Since they are equal, we would deduce that the relationship between the lengths of the arms must be:

$$\frac{2l_B}{c\left(1 - \frac{v^2}{c^2}\right)} = \frac{2l_A}{c\sqrt{1 - \frac{v^2}{c^2}}}$$

$$l_B = l_A\sqrt{1 - \frac{v^2}{c^2}}$$

Therefore, l_B must be the shorter than l_A the greater the velocity v. Yet we have measured two equal lengths for the arms $l_B = l_A = 1$. Lorentz (1853–1928) and independently Fitzgerald (1851–1901) hypothesized that the arm $S-B$ in the direction of motion would contract by the amount that multiplies l_A in the above formula to compensate for the effect of the ether wind on the speed of the light beam. Lorentz thought that, at least within the limits in which no effect of the relative motion of the earth with respect to the ether had been observed, the electromagnetic field should be invariant in passing from the reference, absolute, of the motionless ether, to that of the earth considered inertial, at least in first approximation. Such an invariance is due not only to the contraction of the length and, therefore, of the spatial coordinate in the direction of the motion, but also to the fact that the time evaluated in an inertial system is not the same as that measured in the reference of the ether. We can, for example, think of a space that "reacts" to the motion with respect to the ether with a contraction of the lengths. The speed of light is invariant with respect to the direction (it has the same value independently from the direction of propagation) only in the ether reference. In the references in motion with a certain velocity, instead, it is added to this, compensating its effect with the contraction of the lengths. In passing from one system to another, the measures of space and time change, the first one contracting and the other dilating, as we will see later in greater detail.

Lorentz identified the laws of transformation of space and time, now known by his name. Einstein came to the same result in formulating his theory of special relativity, although he did not share the conclusions about the existence of an immobile ether. Thanks to these conclusions, initially considered at least embarrassing by the scientific community, the ideas of absolute space and time underwent a real upheaval.

Reference

1. Einstein, A. (1931). Maxwell's influence on the development of the conception of physical reality. In *James Clerk Maxwell: A commemoration* (Vol. 1831–1931, pp. 66–73). Cambridge University Press.

4

Thermodynamics

4.1 Time in Classical Thermodynamics (The Arrow of Time)

Thermodynamics was born as the "science of heat" and we could say that it began with the discovery of fire more than a million years ago. It allowed mankind to warm itself, to illuminate the darkness of the night and of the caves, to heat food favouring the absorption of carbohydrates and proteins, to protect itself from predators and so forth. In short, it marked the beginning of the expansion of the human race on earth.

Much later, man began to use fire to operate machines. Heron of Alexandria, with uncertain chronological collocation between the first and third centuries A.D., among his other activities, created a machine that could open doors or, in any case, provide a rotary movement. This consisted of a container containing water with a series of nozzles from which the steam produced by heating it could escape. A kind of whirlwind used today to automatically water gardens. Although the descriptions of the use of steam by Heron had been taken up by Giambattista Della Porta (1535–1615) and before Leonardo Da Vinci (1452–1519) had provided the design of a steam cannon called Architronito,[1] it was not until the second half of 1700 that James Watt (1736–1819) developed his steam engine. At first used to pump

---

[1] It consisted of a barrel with a breech which was heated with a brazier. At a certain moment, water was injected into the barrel, which instantly vaporized and provided the thrust for a projectile. Leonardo attributed the invention of this machine to Archimedes of Syracuse (212 B.C.).

© The Author(s), under exclusive license to Springer Nature Switzerland AG 2022
W. Grassi, *The Challenges of Time*,
https://doi.org/10.1007/978-3-030-94372-1_4

water from mines, around 1810 it had already been produced in 5000 units with applications ranging from steel to textiles. The use of the steam engine spread progressively and powerfully over time, as its characteristics were refined and its further potential was identified. It is enough to remember, among the many, the contribution it brought to the world of transport with steam trains and ships.

The virtuous cycle began whereby the improvement of a technology increasingly implies the need for an in-depth study of its underlying principles. We could say that the initial push was substantially empirical and motivated by practical needs, but it led to fundamental scientific and philosophical consequences.

Without going into too many details, let us recall some scientists who made the first experimental observations in the eighteenth and nineteenth centuries [1]. Among these, Benjamin Thomson, Count of Rumford (1753–1814) and James Prescott Joule (1818–1889) must be mentioned. Drilling cannons to make holes for fuses, Thompson realized that by working with the drill he could generate heat, then considered a fluid called "caloric" contained in bodies, in an amount limited only by the work he did with the drill. It is as if we could draw from a bucket a quantity of water that depends only on how many times we dip the spoon in it.

Joule, with various experimental tests, classical is the one of the rotations of a whirlpool immersed in water, came to the conclusion that heat was a form of energy and found the so-called mechanical equivalent of the calorie. It was also understood, little by little, that the form of energy exchange called "heat" had specific characteristics very different from those of the mechanical work. In fact work can be entirely transformed into heat, but the vice versa is not valid, something that Rumford first realized using the above-mentioned drill. Then, there was Sadi Carnot (1796–1832) who, interested in machines, understood that, since heat always goes spontaneously from warmer bodies to colder ones, it can never be entirely used to transform itself into work: a quota must always be dispersed.

Jean Baptiste Joseph Fourier (1768–1830) had highlighted this "direction of heat" with his studies on thermal conduction, which had demonstrated that the flow of heat always goes from a warmer zone to a colder one. In a bar heated at one end and cooled at the other, thermal energy is always transmitted from the higher temperature to the lower one. The opposite never happens: the phenomenon cannot be reversed. Thus, Fourier put a heavy mortgage on the direction of time: unidirectional and asymmetrical with respect to the phenomenon, unlike what happens in purely mechanical phenomena.

Think about what happens when you start the engine of a car. If you stay in neutral, the water in the radiator heats up, even though no power is transferred to the wheels because you are not moving. The longer you keep the engine running, the higher the water temperature rises to boiling point. All the energy supplied to the engine turns into heat just like the work of Rumford's drill. We could say that you consume gasoline to heat the water in the radiator. In turn, the heat produced is transferred to the surrounding air, and if you turn off the engine, little by little the temperature of the water returns to that of the ambient air. Under these conditions, it never happens that this heat "goes back" to supply energy to the car. While it is true that all the work done by the engine with the movement of the pistons is transformed into heat, it does not happen that at least part of the waste heat goes spontaneously to produce the movement of the pistons.

If, on the other hand, you engage first gear and start, the engine reaches a certain temperature, exchanging heat with the ambient air, but at the same time provides the mechanical work that allows you to move. More precisely, the petrol you consume produces a share of heat useful to cause the movement of the pistons and an ineliminable share that, for the engine to continue to work cyclically, must be dispersed into the environment. This happens in two ways: through heat exchange between water and ambient air and through the escape of exhaust gases from the exhaust. Also, the human body uses a part of the energy supplied by the nourishment to maintain the vital processes and to carry out the usual actions and another one disperses it towards the outside, trying to maintain its temperature around the optimal physiological values.

Again, if you inflate a bicycle with a hand pump, you will feel it heat up as you pump air into the tyre. This heat also dissipates into the environment. All of this suggests that heat is the most likely form of energy, meaning that whatever you do, some of the energy you use is dissipated as heat. But there is more. If you let the tyre of the bicycle deflate by disconnecting the pump or leave the drill bit in the now hot hole, you cannot recover in the form of mechanical work what you have spent in the process, while you leave an indelible trace in the environment. It is said that real macroscopic processes, i.e. of which we have direct experience, are irreversible: it is not possible to go back to the starting point in the same conditions. In fact we already know this, life teaches us continuously. In this sense, there is a before and an after that are not interchangeable. We can also say that there is an instant one, previous, and an instant two, following, and time flows in this direction with a reality that changes progressively, without any possibility of returning exactly as it was before. Events cannot be cancelled, at most their effects can be reduced. This macroscopic non-reversibility makes us say that time flows in only one direction, that of the "arrow of time". All the macroscopic systems we know, be they animals, plants or minerals, seem to obey this rule: they modify themselves by exchanging energy and matter with the surrounding systems and they modify them without the possibility of going back.

For a moment let us think of heat as a boulder placed on top of an inclined plane, say on a smooth, even slope of a hill. What happens if we let it go? We know very well that it falls, perhaps with some bounce, until it reaches a flat area. Can it get back to where it was in the beginning? Here, a physicist would say, "Yes, but not *spontaneously*". And, it is this adverb that makes all the difference. It means that on its own the rock does not even think about going back. But if you work to achieve this, it is a different story. And the heat? If we replace the height of the points on the inclined plane with the temperature (with the largest value at the top and the smallest at the bottom, just as Fourier did when he studied conduction) the form of energy called heat is transmitted *spontaneously* from the highest to the lowest temperature without, however, any rebound. We know it well: if we burn our finger, we put it in cold water and not in the water in which we have just dropped the pasta.

But this also implies something more. The earth is a planet with a large, but not infinite, energy reserve at its centre. It transmits heat towards the space and, little by little, its energy decreases. Therefore, also our planet (and not only) evolves in a certain direction, as well as everything that in the past was thought immutable. Everything changes, but with very different characteristic times, some too long and others too short for man to always have a real perception. Everything, however, in the macroscopic world is born, develops at the expense of resources of energy and matter taken from the surrounding environment, and grows, but at the same time wears out, perishes and dies, becoming confused with it again. This is an indispensable aspect of our daily experience, and it is this that also characterizes our personal destiny. We do not know why, but we know that this is how things are.

Here, Clausius (1822–1888) introduced a quantity called entropy. In a reversible transformation of a system, entropy is the sum of the ratios between the quantity of heat exchanged by the system and the temperatures (absolute thermodynamic in K) of the sources with which it exchanges. If the system exchanges heat towards colder sources (i.e. with temperatures lower than its own) its entropy decreases, while if the sources are warmer, it increases. If the system is adiabatic and impermeable to matter, since there is no possibility of exchange, its entropy remains constant and the system has no possibility of interacting with the rest of the universe.

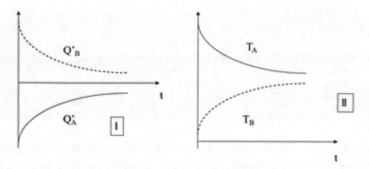

Fig. 4.1 Heat exchanges and temperature trends of bodies

Brief Mathematical Digression

Let us consider two bodies placed inside an adiabatic and rigid container, that is, one that cannot exchange heat, matter or work with other systems and is perfectly empty.

Initially, the two bodies A and B are at two different temperatures T_{A_0} and T_{B_0} with $T_{A_0} > T_{B_0}$. We know from experience that they will exchange instant by instant heat (power) between them, Q'_A given by body A to B and Q'_B received by B. Since bodies can exchange heat only among themselves $(Q'_A + Q'_B = 0)$ we must have $Q'_A = -Q'_B$. According to the usual conventions Q'_A is negative $(Q'_A = -Q')$ because it leaves A and Q'_B is positive $(Q'_B = Q')$ because it is received by B.

Let us see what is the qualitative evolution of these exchanges and temperatures over time, still taken as the independent variable in Fig. 4.1.

Referring to infinitesimal time intervals, dt, the bodies will exchange quantities of heat $dQ'_A = -dQ'_B = -dQ'$. The relative entropy change of the two bodies' system is:

$$dS = \frac{dQ'_A}{T_A} + \frac{dQ'_B}{T_B} = dQ'\left(\frac{1}{T_B} - \frac{1}{T_A}\right) > 0$$

Being $T_A > T_B$. The process will end when the only possible (irreversible) interaction between the two bodies, the thermal one, will cease, that is, when both bodies reach a common temperature. In this case, no entropy increase will be possible any more, while up to this condition, the entropy of the system is always increased by positive elements. In fact, the increase in the system's entropy can be calculated as the sum (integral) of a series of positive elements:

$$\Delta S = S_{\text{final}} - S_{\text{initial}} = \sum_i dS_i = \sum_i dQ_i\left(\frac{1}{T_{B,i}} - \frac{1}{T_{A,i}}\right)$$

As long as the two temperatures do not equal each other, the various dS_i are smaller and smaller positive contributions. Therefore, the value of entropy in such a process always increases until it reaches a maximum value.

It is opportune to underline that the two bodies are macroscopic and therefore constituted by an enormous whole of elementary particles which agitate in various ways inside them and can exchange their consequent energy through the vacuum by radiation, as the sun does with the earth.

Again remember that the only interaction allowed to the bodies is just this kind of heat exchange.

In conclusion, Clausius tells us that there is a quantity (entropy) which is a function of the thermodynamic state. In the case of a system like the two bodies in the previous example, which does not exchange heat or mass with others, this always increases due to a non-reversible transformation. If we looked at it with an infrared viewer, in the dark, and if the bodies had the same dimensions and the same shape (e.g. two spheres of the same radius) how would we distinguish body *A* from body *B*, having only information about their temperatures? At the beginning, body *A* is hotter than *B*. At this instant, we can say, for example, that body *A* is on the left and *B* is on the right. At the end of the process, when the two bodies have the same temperature, i.e. they are in thermal equilibrium, no one will be able to say where *A* is and where *B* is, bearing in mind that someone could have changed their reciprocal position. *A* and *B* are, therefore, indistinguishable. You go from a situation where they have a certain order and can be easily distinguished, to one where this does not happen. Kind of like what happens when you walk into a newly remodelled hotel room. Everything is in its place. Then you start to unpack the bed, fluff the pillow, move a chair to take off your shoes, open the curtains and so on. Gradually you unclutter and, after a while, you are looking for your glasses or the map you found on the bedside table and took a quick look at. You find it more difficult to identify the objects that get more and more confused with the others. If you have a child in the house, you know even better what I mean. And you also know how much energy (sweat is a manifestation of this) it costs you to tidy up, since unfortunately you do not have any Mary Poppins to do it.

Moreover, if the exchange of heat and the consequent variation of temperature are the only manifestations that you can perceive of the evolution, or if you prefer of the "vitality", of the two bodies, they cease at the moment in which thermal equilibrium is reached. From this point on, they are no longer vital. They interact with each other only as long as there is a certain unbalance, precisely the difference between the temperatures. In relation to them, we could say that their evolutionary time begins when the maximum difference in temperature exists and ends at the moment in which it is annulled.

The things, however, are still more complicated, as we will see in a little while, from the fact that any macroscopic system is generally constituted by subsystems interacting among them (think about the human organism consti- tuted by various organs) and interacts with others in variously complex ways. Let us see, referring to an apparatus as simple as the pendulum, how the situ- ation changes in passing from the ideal world of the Newtonian mechanics to the one in which we live, strongly irreversible.

If we consider a perfect pendulum in vacuum, we know that, if at a given moment we move it from its equilibrium position, in which the rod at the end of which a weight is placed is perfectly vertical with the weight at the bottom, it begins to oscillate and continues to do so indefinitely, just as Galileo said. If we only place it in air, which is a viscous fluid, the pendulum slows down and after a while stops. Precisely because of the viscosity, a part of the energy of the pendulum is dissipated in heat. Even if we know perfectly the mass of the pendulum and its length, we can say that it will stop after a given interval, always the same? The answer is: yes it will stop and no it will not always stop after the same interval of time, because there are many variables of which we do not know the trend. The temperatures of the air and of the pendulum can change and modify the characteristics of both (dimensions of the pendulum and viscosity of the air), even small perturbations of the flow of the surrounding air can disturb the motion, the hinge on which the rod is fixed wears out and can dissipate more energy in friction and so on. So, even if we know that the pendulum will stop, we are not able to say that it will always take the same time because things can happen that are not always predictable.

The fact that, however, the pendulum stops or that the two bodies of before are brought to the same temperature already identifies the existence of what elsewhere takes the name of attractor. In the previous cases, it is a point, iden- tified by a value of the temperature or of the angle of the pendulum, null with respect to the vertical, towards which the phenomenon is attracted and ends. In a general sense, the attractor is a set of points towards which the system evolves after a long enough time. Very trivially anyone who goes to work leaves home with an advance that generally assures him to arrive on time, but certainly he will not be able to tell you the exact minute he will arrive. The interactions with the bus route of the various systems that can condi- tion the bus times (from traffic, to the number of passengers at the stops, to the functioning of traffic lights and so on) can have more or less importance. These are variables that generally constitute disturbances of limited impor- tance. So, you know that you will arrive at the office at a certain time, and you take the necessary measures not to be late. But if there is an accident, a sudden strike or a breakdown on the road, i.e. if the above disturbances are amplified, the whole working day can be blown. If you then have a career-decisive appointment that day, your working future can change. Therefore, if the system under scrutiny (e.g. the bus) interacts with others, its behavior can be variously modified and even disrupted. The disruption and uncertainty in the unfolding of your day increases and can do so to a great extent. We are no longer in that "protected" system constituted by the two bodies that could only exchange heat with each other and whose fate you could easily predict. In more complex conditions, the dynamics of events is no longer so simple. Everything depends on allowing the system to be subjected to the actions of others increasing the possibility of external influences. From a system such as the two bodies, which could only exchange heat internally, we have moved on to systems that are free to relate to others in ways that are often not knowable and that contribute to accentuate imbalances. We have changed the charac-teristics of the "frontier" of the system, that is, of the surface that delimits it. As we will see, living organisms have borders that are permeable to the passage of energy and matter and this is why they are called open systems; they are in disequilibrium with respect to the environment that surrounds them, but this does not mean that they are disordered, on the contrary. Man is in profound disequilibrium with his environment: he has a different temper-ature, he moves and behaves in certain ways, overcoming or avoiding the difficulties it poses (he tills the soil, sows the seeds, lights a fire, goes hunting, builds houses, etc.) and even drastically modifying it. How does he do all this? At the base, he takes in energy through food and expels substances that are not useful to him. In his daily activity, similar to a thermal machine, he takes energy from nature and gives it back waste and heat. It can continue to survive as a human being as long as it can carry out these operations and as long as all its internal organs and their irreversible interactions function properly. So, it goes for animals, trees, and so many phenomena that occur in nature. These are what Ilya Prigogine (1917–2003) called dissipative struc-tures, i.e. structures that, although deeply irreversible, in disequilibrium with the surrounding environment and the site of irreversible phenomena them-selves, are nevertheless ordered. Concluding, therefore, that irreversibility can give rise to ordered structures that subsist as long as they can exchange resources with the surrounding universe. However, they often influence each other in an unpredictable and sometimes unexpected way.

Now it is necessary to emphasize a point that is neither trivial nor negli-gible. Classical thermodynamics as well as many classical sciences that deal with matter as a continuum examines systems consisting of an extremely large number of particles.[2] Thus, temperature is an index of their agitation and hence their average energy of motion, density provides a measure of mass per unit volume, etc. Heat flow involves the transfer of energy by a large number of particles. This is why we have often used the adjective macroscopic earlier. Whenever we consider a thermodynamic quantity, we are referring to extremely large numbers of elementary units. So also when we make statistics to have predictive possibilities, we have to refer to "big numbers".

We can say that the probability of getting heads or tails by flipping a coin is 50% only if we make a large number of flips. Otherwise, the result can be any. Just as before deciding how long it takes you to get to the office from home, you have done a number of tests.

Continuous and Macroscopic

We have repeatedly mentioned the term "macroscopic", and we have done the same with the term "continuous". In one case, we spoke of systems containing a very large number of elements, in the other, for example, of the extension of space without interruptions. These two approaches are not unrelated. Try to look at a fabric. Visualizing what we are saying is more or less easy depending on whether you are in the presence of a wool sweater, a carpet or a silk shirt. Sometimes, you can see it with the naked eye, simply by getting a little closer, and other times, you need a magnifying glass or even a microscope. In any case, you always end up seeing the same thing, depending on how close you get (optically) to the characteristic dimensions of the weave. From a distance, the fabric looks like a sheet of paper or a liquid surface. You can move over it with very small movements and you encounter the same kind of matter everywhere. Now look at the woollen sweater your wife has made for you. It does not take you long to notice the presence of interwoven fibers enclosing air. Of course, you cannot see the deep structure of the fibers and the air. You would need instruments far more sophisticated than the eye. However, you have caught the discontinuity of wool fibres and air, which you would not have been able to see from a distance. Just as well you distinguish billiard balls during a game because they are quite large compared to the surface of the billiard table and to your visual capabilities. At the same time, you are unable to distinguish the texture of the billiard table fabric on which they are moving. Few objects large enough can be seen well, and you understand that they are separated from the air, which, in turn, contains an enormity of tiny elements.

This is the macroscopic sensation that they give and as such we treat them, applying a sort of physics of large numbers and studying their overall behaviour referring to quantities that interpret the overall behaviour, such as pressure and temperature. The differential mathematical analysis we use is also a consequence of this point of view, and we use it to the full for the study of

---

[2] Keep in mind that the number of elementary particles (Avogadro's number) in a mole of substance is 6.0221409×10^{23} and that, for example, the mass of a mole of water is about 18 g.

the theory of the elasticity of solids, the motion of fluids and the propagation of electromagnetic fields.

When this approach cannot be valid we have to resort to different methods of investigation which, we should not be surprised, if they can lead us to different conclusions.

The question is, then, in the case of a very small number of particles does what we have said so far still apply? While the principle of conservation of energy is still valid, something must be said about the second principle and about the behaviour of entropy and, therefore, of the direction of time. We should not be surprised by this because exquisitely classical thermodynamic concepts have a statistical value and, consequently, apply to sets formed by a large number of elements. For limited numbers and with a reduced number of degrees of freedom, they can even lose their meaning. In the words of Mandò and Calamai [2]: "the *first principle is nothing else than the principle of conservation of energy and, as such, it is applicable to a system of few particles, or even only to a single particle; we could say that, passing in the molecular world, the first principle reveals its true character...non-thermodynamic; the concept of heat quantity is a thermodynamic concept, but it represents only a synthetic macroscopic means of summarizing certain modes of energy exchange which only the enormous complication due to the large number of particles prevents us from following individually, particle by particle; but if we did so, we would find the principle verified for each individual particle, for each individual collision*". As said, things are quite different for the second principle of thermodynamics, the one that for the two bodies of before, anyway composed by a huge number of particles, implied an increase of entropy. Let us refer to a classic experiment performed by Joule related to the free expansion of a gas.

Let us suppose we have two identical containers connected by a pipe with a valve or a separation septum. In one of them, for example the one on the left, is placed gas under certain conditions of pressure and temperature. In the one on the right, a vacuum is created. When the valve is opened, the gas will fill both containers at a lower pressure than the one on the left. The gas will move from left to right until the pressure in the two containers will be the same. Can it happen at this point that the gas contained in the right-hand container returns entirely to the left-hand one spontaneously? Maybe some molecules will jump randomly from one to the other with small *fluctuations*, but no one has ever seen a room empty of air once he has opened a window. The process is irreversible, and once the pressure has equalized in the two containers, everything remains macroscopically still. The molecules are so many, and basically, they are arranged half in one and half in the other container. It does not matter if there are a few thousand more molecules on the left and fewer on the right. But if the number of molecules is reduced, a lot the matter changes. If we behave in a very drastic way and we put in the left container two molecules and in the right one we leave the vacuum, where will we find them once the valve is opened? Let us help ourselves again with the words of Mandò [2]: "...*who will be able to prevent one of the two molecules, in its disorderly motion, from entering the separation duct and going to keep company with the other, or who will oblige the other, once the first has crossed the door of the house, to rush to the exit so as not to find itself together with the other?*" It will no longer be true that molecules must go one into one vessel and the other into the other. If we increase the number of molecules present in the left container up to one or two moles, we still have the possibility that the second principle is violated, but the probability is really very low, or that fluctuations occur, in any case of minimal entity, with not negligible probability. Continuing with an analysis much more careful than this, we come to the conclusion that the second principle has only a statistical value. We would say that it is a certainty only for systems of macroscopic dimensions.

We can look at what we have discussed also with a slightly different point of view. When the left container contains, at the beginning, a macroscopic quantity of gas, the average free path[3] of the particles is reduced, the interactions among the particles are frequent, the pressure inside the container has a precise value and the unbalance between the pressure of the left and the right container is well defined and quite high. If instead there are only two particles, the average free path is large, the interactions are few and the aforesaid unbalance, which is at the origin of the irreversibility of the phenomenon, is very small.

Think about what happens in a bus that has just left the depot when it makes the first stop. If the number of waiting passengers is high, they will tend to distribute themselves in the bus with a certain uniformity in order to maximize the reciprocal distance (average free walk) so as not to bother each other. Two friends, who want to stay close to each other to continue the discussion started at the bus stop, will have their work cut out for them. But if there are only two passengers at the stop, it is difficult to predict whether they will sit close together or far apart. If, during the journey, you pass through an interesting part of the city, it may be that they come closer to see it better, perhaps from the rear window, and then move away again. The same thing does not happen when the bus is full. Therefore, it does not really make sense to say that the two passengers will spread out evenly in the bus. If the

---

[3] The mean free path is the average distance travelled by a particle between two successive collisions, i.e. before it collides with another particle.

number of passengers is greater, but not so great as to fill all the space, we would notice that there may be occasional crowding (fluctuations), perhaps near the entrance or exit when arriving at stops. On average, however, people tend not to bother each other and therefore remain as evenly distributed as possible.

In one case, the direction of the event is well specified. If we look at the watch we have on our wrist, we say that first all the particles are in one container and, after the valve is opened, they will occupy all the available space, and therefore both containers. In the case of two particles, they can be anywhere. According to our clock, while before the opening of the valve, they were both on one side, afterwards (i.e. when the clock strikes a few more seconds) they can still be on the same side or on different sides. After observing them for a while, we can reasonably say that they spend on average the same amount of time (always according to the clock) together, in the left or right container, or separated.

But if we do not have clocks and we want to say how time flows, what conclusion do we come to? For the macroscopic situation, time flows in the same direction that most particles go. The past corresponds to having them all in the left tank and the present to having them evenly distributed in both. Without further intervention things will stay that way. A real future, that the system will spontaneously create itself, does not exist.

If there are only two particles and we observe the system once the valve is open, how do we distinguish between past present and future? We certainly cannot say that the past coincides with both particles in the left tank, the present with one particle on each side and the future with both on the right. In fact what we would read on the clock would have nothing to do with this conclusion.

The transition from a macroscopic to a microscopic view was due to the rise of statistical physics. It had already been used by Daniel Bernoulli (1700–1782) to explain the pressure exerted on the walls of a container through the collisions of the molecules of a gas contained in it. In other words, he attributed a macroscopic quantity like pressure to the motion of many microscopic components. This kinetic theory (of the motion of elementary particles) was subsequently developed in particular by two eminent scientists: James Clerk Maxwell (1833–1879) and Ludwig Boltzmann (1844–1906). According to this, in particular, heat was attributable to the chaotic motion of molecules and pressure and temperature were phenomenological manifestations of this motion. In other words, they are the manifestation in the macroscopic world, i.e., in the world that is immediately "visible" or perceptible to us, of the agitation of myriad particles.

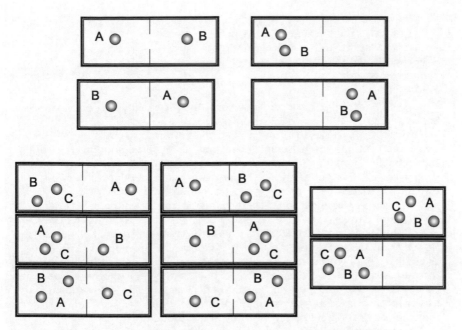

Fig. 4.2 Possible particle arrangements in two communicating reservoirs

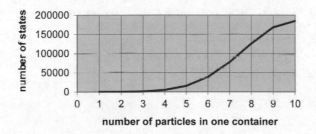

Fig. 4.3 Number of possible states as a function of the number of particles present

Brief Mathematical Digression

Let us consider again the case of the two tanks and the free expansion of the left gas to the right after the opening of the partition septum. Let us go step by step following the various events in relation to the number of elementary particles present.

A single molecule can carry itself either to the left or to the right. The probability that it is in either of the two containers is ½ or 50%. This is the number of possible microscopic states. If there are 2 particles (shown in the figure for clarity as *A* and *B*, but remember that they are identical), they can arrange themselves in four different ways, as shown in Fig. 4.2. The possible arrangements are:

- Particle *A* to the left and *B* to the right;
- Particle *B* on the left and *A* on the right;
- *A* and *B* together on the left;
- *A* and *B* together on the right.

The system has four possible configurations. The probability that both particles are on the left is $(1/2)^2 = 1/4$, 25%.

So, for three particles this same probability is reduced to $(1/2)^3 = 1/8$, 12.5%.

We can continue the reasoning. For a number, *N*, large enough of particles the probability of finding them all in the same container (left or right) is about given by $(1/2)^N$. For 100 particles, this probability is around 10^{-31}. For a mole of gas, $N = 6.0221409 \times 10^{23}$, it is practically non-existent.

If we already consider only three particles, the opening of the separation septum between the two vessels determines the passage of the possible microstates of the gas from a certain number to a larger one. Initially, all particles were on the left. After the opening the possible arrangements are more. While there is only one way to have all the molecules on the left, there are three ways to have two on the left and one on the right and three ways to have two on the right and one on the left. So, there are six possibilities to have two particles at the same time in the same container, which means that there are six possible microstates for this configuration.

If there were 20 molecules in all, the number of possible microstates as a function of those in a container increases as shown in Fig. 4.3. It increases until there are 10 particles on one side and, of course, ten on the other.

It was Boltzmann who understood that in an isolated system, such as the set of two reservoirs, entropy tends to increase because there are many more possibilities to have high entropy than low entropy. He linked entropy, *S*, to the number of possible microstates, *W*, according to the relation:

$$S = k \ln W$$

$$k = 1.3864\ldots \times 10^{-23} \text{ J/K}$$

The increase of entropy in an irreversible phenomenon was, therefore, referred to the change in the mechanical characteristics of a system of interacting particles, which go from a more to a less ordered situation. This highlights the statistical character of the second principle of thermodynamics, that is, its link to a statistical interpretation rather than its absolute validity.

Using also the relation written in the previous box, we therefore say that the greater the number of ways in which an event can occur, the greater the probability of its occurrence. In addition, statistical physics, in particular with Boltzmann, underlines a distinction between a microscopic and a macroscopic vision of the physical knowledge of nature. In addition, the second principle seems to define a certain arrow of time for systems consisting of a large number of elements. For example, the macroscopic state of stillness of a system, directly perceivable by us, depends on a movement, so to speak invisible, of an infinity of molecules. All this makes a statistical approach essential even for the study of a very small part of matter. Such a conclusion costs Boltzmann the tenacious opposition of a number of scholars, including Ernst Mach (1838–1916) and Wilhelm Ostwald (1853–1932), who advocated a purely phenomenological, or empirical, approach to physics and rejected atomism. In contrast, Boltzmann's attitude emphasized the existence of a kind of dichotomy between the sensible (phenomenological) approach and knowledge that some would call conceptual. Instead of simply "recording" the data provided by experience, phenomenology should increasingly push us beyond experience. Subsequent developments in physics have demonstrated the validity of this position, for which Boltzmann had fought, often alone during his stormy life.

We experience the movement of things towards situations in which there is a greater number of configurations even in a macroscopic way, therefore directly observable with our senses. If we drop a plate on the floor, it breaks or, as they say, it breaks into a thousand pieces. The plate is a unique, precise configuration of the various pieces from which it is composed. It is an extremely orderly whole with a well-defined pattern. When it breaks, it takes on another, and the number of parts into which it is divided is decidedly much greater than one, while the design is broken up into incoherent shapes. Though dropping it from the same height, the number of pieces and their shape is different each time. Can it break twice in the same way? It can happen, but it is very unlikely because the number of possibilities it has is quite high. So, try throwing up a deck of cards that you have patiently sorted by suit and increasing values. What is the probability that they will fall back forming a deck still sorted as before? Practically none. You would have to have infinite patience to wait for that to happen, it has to choose from too many other possibilities. Spontaneously certain things do not happen in real life that is limited in time.

What can we conclude from what we have said? In two words, we can say that in systems formed by very many elements[4] the direction of time goes towards the achievement of thermal equilibrium, or of maximum indistinguishability or, again, of maximum disorder. This, however, remains an idea of the macroscopic world, even if, as we will see in the next paragraph, things are a bit more complicated. In any case, it remains what Russel said that time is different for "*bodies of large dimensions*" and therefore composed of many

---

[4] Even in the case of two bodies at different temperatures in an adiabatic environment, we are referring to systems composed of huge numbers of molecules and that is why we can characterize them by their temperature.

elements with respect to single particles, even if Russel refers to very small bodies moving with velocities close to that of light.

4.2   Time and the Thermodynamics of Dissipative Structures

The system of the two recipients that can only exchange particles between them, but cannot have any other exchange of matter or energy with the outside, that is with what is beyond their physical walls has, therefore, only one possibility. In it only the mass of gas present can be redistributed. This happens in the direction of the growth of disorder and with the increase of the value of a thermodynamic index called entropy. This conclusion leads us to say that a macroscopic system, which is therefore made up of a very large number of elements, has a very high probability of reaching a situation of final stasis, after which it no longer gives "signs of life". We cannot a priori exclude that the system spontaneously returns to the initial conditions, but the probability is so low that it has never been experimentally verified. Different is the matter for low numbers of elements that compose the system, and the previous considerations are no longer valid. On the other hand anyone can tell you that the behavior of a huge crowd is different from that of a group of three people. So no wonder.

A system like the one considered has, however, very few possibilities to evolve, not having macroscopic interactions with other systems, even if the number of interactions among the single elements is enormous. In it exists a before and an after with an arrow of the time univocally directed, if we exclude happenings of substantially null probability, above all on human times. Now, referring to an example already discussed in another text [3], let us consider very common physical events such as the movement of air in a room where you have turned on a heat source (from a stove to a radiator) or the formation of a breeze in nature. If we turn on a stove in the kitchen, whether gas or electric, at some point, the air above it starts to rise and, if there is one, as it should, it goes up the hood. Let us see what happened, referring for simplicity to an electric stove. Before ignition, the air is evenly distributed and the molecules move randomly. Once the hotplate is turned on, a temperature difference is created between its surface and the air near it. The heat goes from the hotter air to the nearest gas molecules, which start to agitate more intensely, without yet triggering a macroscopic motion. When the temperature of the plate reaches a certain value (it is hot enough), an upward macroscopic motion begins, which continues as long as the plate remains hot. An imbalance has been created between the temperature of the air near the plate and the rest, strong enough to give rise to this updraft. Macroscopic motion exists only if this temperature difference is maintained, that is, if energy is provided that can overcome the friction caused by the motion of the molecules, which would tend to stop it. The molecules involved in the fluid current, caused by a local unbalance in the room (temperature difference) can continue to follow the flow or return to the initial situation of "macroscopic calm" depending on whether we keep the plate hot or turn it off. We could also say that, by turning on the hotplate, we cause a perturbation and, therefore, a fluctuation in the local behaviour of the molecules. If we continue to supply energy this fluctuation is enhanced and can remain in a stable configuration, with a flux that depends on the amount of energy supplied. If we turn off the plate, for example just after the macroscopic motion has been established, it is quickly exhausted and everything returns as before. The energy we supply is the element that determines the amplification of the fluctuation and its subsequent stable permanence. To make it cease, it is enough to stop supplying it by interrupting this supply. The breezes of land and sea do something similar. When the sun rises, the earth begins to warm up more rapidly than the sea. At a certain point, the air on the earth, as it warms up, begins to rise, drawing air from the sea, which moves parallel to the ground. The process continues until the temperature difference between the ground and the sea becomes too small. The movement stops, to start again in the evening in the opposite direction, when the sea is warmer than the land. We stress again that for all this to happen it is necessary that there is a contribution of energy (electric plate or sun) and a continuous input of new air (mass exchange from the surrounding areas). The fluid current system is said to be an open system.

Figure 4.4 qualitatively describes the behavior of the molecules. Starting to heat the plate increases their agitation without determining any macroscopic motion (a). When the plate is hot enough, they give rise to an air current with its own macroscopic dynamic structure (b).

The occurrence of a local unbalance can then lead to two different effects. In the case of a system like the one of the two reservoirs, the unbalance due to the strong initial inhomogeneity in the distribution of the particles, if these are numerous enough, causes their diffusion and a consequent uniform distribution in the system formed by the reservoirs. Since there is no contribution of energy or further matter, as we said before, the system no longer has a "macroscopic future". There may be some internal fluctuations of groups of molecules, but nothing relevant to the effects of any overall change.

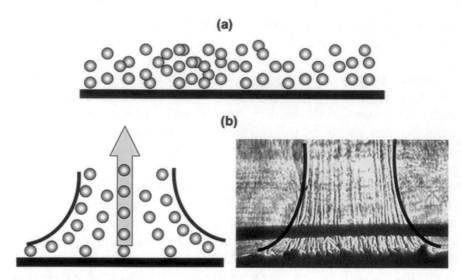

Fig. 4.4 Particle motion. At the top, they move disorderly on short paths and the air is macroscopically stationary. When the plate is hot enough, the macroscopic convective motions begin at the bottom

If, however, the air-quiet system of a room is locally perturbed and this perturbation is sustained, as in the case of the plate, which is still a small object compared to the size of the entire room system, a macroscopic structure such as the air current is created and can remain stable as such.

It originates, then, something new with its own order, in absolute disequilibrium with the surrounding environment. Therefore, a new type of structure is born. It is no longer the state of macroscopically still air in the room (which can still be so, far from the plate), however, determined by a disordered motion of the molecules of which we are not aware, but an overall dynamic order that we can directly perceive. We ourselves are an example of ordered systems able to maintain our structure stable as long as we have the possibility to interact with the environment and the subsystems (organs, etc.) that compose us function and interact with each other in an appropriate way.

Prigogine and Stengers say [4]: "*The equilibrium structures can be seen as the result of a statistical compensation of the activity of the multitude of microscopic elements (atoms, molecules). They are, by definition, devoid of macroscopic activity, inert at the global level. In a certain sense they are also immortal; once formed, they can be isolated and maintain themselves indefinitely without any further need for exchange with the environment. Now when we study a cell, or a city, the situation is quite different. These systems are not only open, but they live only because they are open. They feed on the flow of matter and energy that comes to them from the outside world. It is not possible for a city or a living cell to evolve to a new compensation, to a balance between incoming and outgoing flows. We can, if we want, isolate a crystal, but the city and the cell separated from their environment, quickly die. They are an integral part of the world that nourishes them, they constitute a sort of incarnation, local and particular, of those flows that they ceaselessly transform"*.

If we are able to observe the motion of the air, we realize that it assumes different characteristics depending on how strong the interaction with the plate and the surrounding air is. The motion of the so-called fluid particles (which are very small, but contain very high numbers of elementary particles such as atoms and molecules) takes on different types of order, giving the current different peculiarities. When the plate is progressively heated, at first they move in a laminar fashion, consisting of many rows of juxtaposed particles moving upwards. Then, when the plate is hotter, they pass from a situation in which they seem to want to decide whether to keep the same type of motion or change it: transition period. By heating a little more, the motion changes and the structure changes: we pass to turbulence.

A similar phenomenon, but much more easily observed, is that of the jet of water coming out of a tap. If you open it a little, maintaining a small flow of water, you see a very regular and transparent structure. Opening it more, after a transition phase, the jet loses its transparency until it takes on a milky appearance, also related to the air that is trapped in it. We have also passed here to the turbulent regime. Even if we are not able to distinguish it with the naked eye, at this point, from a structure with liquid layers that move parallel, we have passed to a structure in which vortices of different size and energy are formed with regularity. The new structure is still ordered, but with a different order and needs more energy (flow) to be maintained. The system is, therefore, able to self-organize in a new configuration that is also, beyond appearances, ordered. Self-organization is a peculiar property of systems far from equilibrium. Even the characteristic spatial and temporal scale lengths have changed. Again using the words of Prigogine and Stengers: *"...The parameters that describe them are macroscopic, they are not of the order of 10^{-8} cm, like the distances between the molecules of a crystal, but of the order of the centimetre. Similarly, the time scales are different, they do not correspond to molecular time (the period of vibration of an individual molecule is about 10^{-15} seconds) but to macroscopic times: seconds, minutes, or hours"*.

Earlier, we saw that in normal conditions close to equilibrium, as in the case of the two gas reservoirs, a thermodynamic system undergoes an evolution towards a stable equilibrium condition, reaching a maximum of entropy. After that any further change is impossible. In other conditions, if the system is open and can therefore interact with the outside world for the exchange of energy and matter, it can stabilize in a dynamic state far from equilibrium. In the case of fluid motions, for example, millions of molecules move according to defined collective motions. We are therefore in the presence of a real organized structure, born spontaneously from a homogeneous state. According to external contributions, such a structure can also change into another one, however, always organized.

If, instead of one, we have two identical hot plates in the room, as in Fig. 4.5, each with its own exhaust hood, things can get complicated depending on the possibility they have of interacting with each other.

If the room is large enough and the plates are far apart, each one behaves as if it were alone (see the configuration at the top of Fig. 4.5). If, on the other hand, they are close to each other, they interfere and it may happen that, depending on the conditions of the hoods, some difference in temperature of the plates, the fact that one has been turned on a little earlier than the other and so on, the air flow of both may exit from one hood only. In the figure, it is considered the case in which all the hot gas exits from hood *A*, while hood *B* does not work. Of course, the symmetrical situation can occur. At a certain moment something not well definable (any disturbance) determines the exit

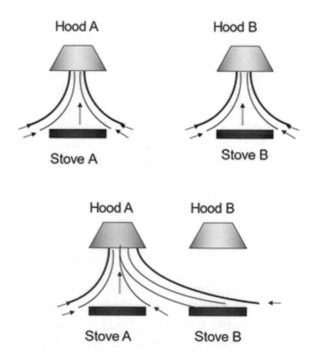

Fig. 4.5 Air heated by the plates exits from the hood above the respective plate, at the top. In case of close plates and/or due to disturbances, both air flows can affect only one hood

conditions of the hot air.[5] It seems that the airflow system from one of the plates decides whether it should go straight up or to the right or, in the other case, to the left. This depends on what happened immediately before the start of the movement to the system under examination and to the other (second plate) that is related to it. Or even if, once it has started, any perturbation occurs. We could say that it is the "recent history" that decides. What counts is not so much time as a univocal direction, but the time that, instant by instant, brings "fresh" information to the system and that, starting from air that is heated on two distinct plates, ends up forming two separate currents or a single one directed to the left or to the right. Time is therefore called upon, instant by instant, to give a form and an organization to the structure that is being created.

We pass from a time that "coldly" imposes a direction, i.e. a law, on natural phenomena to one that participates directly in the processes of formation, choice and evolution of phenomena. The only thing we could have foreseen as a function of the initial conditions, in classical terms, is that due to the heating of the air on the plates it would have moved upwards, becoming lighter.

Here again we try to trivialize with an example. Let us imagine that we are in a terminal station like Roma Termini. We are waiting for the train in which we have a seat in carriage number 7. What happens if we are the only passenger and the arriving train is empty? As soon as it stops on the platform, we move from our position and go straight to the door of the carriage. The path we take is straight, starting from our initial position and going, practically without change directly to the point of arrival. In real life, it never goes like that. There are passengers who wait with us and others who leave the train as soon as it arrives. Normally, the path we take is never that straight and we often have to modify its direction locally because of the many interactions with others. The changes we have to make locally to the path do not give a damn about knowing what the exact point we started from is. If we even fix our attention on the people who have left the train and are coming towards us, how do we deal with them? Whenever they are in front of us, we are not sure where to go. We dodge them by passing from the right or the left. Our future trajectory is not so sure and we look both ways to see what lies ahead. We can say that the trajectory is not as stable as before. From time to time, we are presented with two or more possibilities to choose from.

---

[5] From a practical point of view, such cases have occurred. Think of when the boiler (unsealed chamber) in an apartment was placed in the kitchen, where the stove is also located. Both the boiler and the stove used the air in the room for combustion. In doing so they could significantly interfere with each other to the extent that the fumes from the boiler were drawn into the kitchen. The technical regulations had to exclude this possibility absolutely.

The path we can follow "forks", and it is up to us to choose where to go. Finally, we decide for the right and, maybe a little later we run into a group of tourists that makes it more difficult for us to pass, but we had not seen them because they were hidden by a cart. This can even force us to pass on the other side of the sidewalk and maybe make us meet a friend that we had not seen for a long time. It went well, think instead if it had forced us to stop with one of those boring, long-winded acquaintances that force us to suffer the story of their adventures. In essence, the path that we do is "updated" continuously and depends from time to time on the momentary conditions, regardless of the initial ones. While in the first case, we knew precisely what was the most convenient route and just put one foot in front of the other to reach our destination, now, we must instantly change the path reorganizing it according to the information we receive from the surrounding situation.

Basically, the other passengers force us not to be able to distract ourselves too much by looking at our mobile phones, which we could have easily done if we were alone. You could say that we resemble one of the gas molecules we were talking about earlier, subject to the random collisions of the others. But in this case, we do not expect any Boltzmann equilibrium to be reached. The final point of arrival corresponds to the emptying of the platform of the track, that is, when the structure formed by the stream of passengers descending from the train and those getting on ceases to exist, in this case due to the absence of other people. That "mass flow" that previously fed the structure of individuals in movement is missing.

In one case, time is a simple and independent variable of the phenomenon. It is enough to know the starting point, initial condition and arrival point to have all the information we need. We notice its passing only if we look at the clock. In the concrete case, when we are no longer alone; each instant brings new information and implies consequent decisions and actions, modifying the process as it unfolds. It can be said that this leads to a vision of time as the bearer of ever new information which builds, in most cases, the history of the evolution of systems, conditioning their choices. Time conditions, therefore, the processes sometimes creating autonomous ones and, however, leading them to evolve towards situations not always predictable through the knowledge of the initial conditions. According to some, and in particular Ilya Prigogine (1917–2003) [4], nature is characterized by a history that is no longer necessarily anchored to the necessity of laws, but is "free" in its creative force of processes. So writes Valotta in an article commemorating the centenary of the birth of Prigogine [5] "... *In this turbulent scenario, at the end of a troubled path, in the course of which the theoretical-epistemological crisis that invested classical physics played a central role, there is the figure of Ilya Prigogine, whose explorations on thermodynamic phenomena, far from thermal equilibrium, the so-called dissipative structures, have shown that temporal irreversibility constitutes the irreducible existential dimension of cosmic evolution.*

As Bocchi and Ceruti have written on this subject: The contribution of Ilya Prigogine to the study of the problem of time… has removed the reasons for any radical separation between science and time… To the illusion time of mechanics … and to the degradation time of classical thermodynamics, we now add the possibility of conceiving a time of creation whose symbol is the instability of systems that can pass from one structure to another precisely because they are unstable.

The pivot around which Prigogine's speculation revolves is, therefore, the idea that time, far from being that aseptic "geometric parameter", aimed at guaranteeing the definitiveness of our gnoseological acquisitions, possesses a historical-factual concreteness, in that "it presides over the transformations of matter at the same time in which it is conditioned by them". It is in the attitude that modern scientists have adopted towards the idea of time that Prigogine identifies, in particular, that reductionist fallacy[6] that has led our civilization to deny the unpredictability of nature. Dramatically simplified to the point of negation, time was, in fact, considered by classical science as a repetitive event, always identical to itself; this could only lead to a robotic vision of nature. The impulse given by Prigogine to the start of the non-linear thermodynamics, has given a hard blow to this mechanistic conception because it has shown that it is only in conditions close to turbulence that matter can acquire new properties compatible with life. Since, moreover, every natural event interacts with the surrounding environment, developing in a specific temporal direction whose evolution cannot in any way be foreseen, this means that it is not possible to go back to an objective external cause without abstracting the object from its context, in peace with the omniscient demon hypothesized by Laplace, who, as Prigogine and Stengers state, "despite his calculating virtuosity, is not "closer" than us to a deterministic description".

---

[6] Reductionism is the position that studies a phenomenon by subdividing it into many separate, simpler sub-phenomena. From the study of these sub-phenomena it is claimed to go back to the more complex main phenomenon. We have to be very careful because with this procedure we tend to eliminate the interaction effect among sub-phenomena. If we want to study the behaviour of a group of people, it is not sufficient to study the single individuals separately. The knowledge of the single is necessary, but if we don't know the interactions among the individuals, also linking them to the situations, we can't predict the behaviour of the group. As in many other cases (think of wanting to study the behaviour of the human body by contenting oneself with the anatomy of the various organs) it is not true that the behaviour of a whole (group) is simply the sum of the behaviours of the elements (single individuals). The so-called superposition of the effects is not valid, except, at most, in the very first approximation, because the phenomena are not linear.

References

1. Grassi, W. (2017). *History of heat and cold—Energy and life: Everything is transformed*. Hoepli. (in Italian).
2. Mandò, M., & Calamai, G. (1968). *Lectures on general physics* (pp. 458–460). Libreria Universitaria (in Italian).
3. Grassi, W. (2018). *Butterflies and hurricanes—Complexity: The theory that rules the world*. Hoepli. (in Italian).
4. Prigogine, I., & Stengers, I. (2007). *The new alliance—Methamorphosis of science* (p. 134). G. Einaudi s.p.a. (in Italian) or in the original language: (1983). *La nouvelle alliance – Metamorphose de la science*. Gallimard NRF.
5. Valotta, B. (2017). Ilya Prigogine's revolution of time. *Epekeina, 8*(1), 1–14 (in Italian). Ontology and Cognition, ISSN: 2281–3209; https://doi.org/10.7408/epkn.1. Published online by: CRF - Centro Internazionale per la Ricerca Filosofica - Palermo (Italy). www.ricercalosoca.it/epekeina

5

The Time of Relativity

5.1   There is no One Faster Than You

With Galileo, we learned that saying that we are stationary or in motion has
no meaning if we do not specify what we are doing. If we are reading a book
sitting in a train that is taking us to Rome, we say that we are stationary,
unlike the conductor who is moving to check tickets. The passenger in a
car driving along the road that runs parallel to the railway will also have the
conviction that he is sitting quietly in the car enjoying the panorama that
seems to be coming towards him and then moving away. We too have the
same impression if we look out of the window, away from the book. Not
only that, for us the car is moving slower or faster than we are, unless it is
going at the same speed as the train. It all sounds like a big mess, but only
because we have improperly used the terms stopped and in motion.

Let us try to be more precise. In going to Rome, the train is moving with
respect to the ground, and we are content to take this as a fixed reference,
even though the earth is moving rapidly about its axis and the sun. At this
point, we can say that, if sitting and reading, we are stationary with respect
to the train, while the controller is moving with respect to it. Automobiles
travelling down the road are moving relative to both the train and the ground.
Moreover, if the train has a speed of 100 km/h (of course with respect to the
ground) and a car has a speed of 80 km/h, we see it moving away, after having
passed it, with a speed of 20 km/h. Finally, we have put some order into it. If
we take the ground as a reference, the train moves at 100 km/h and the car
at 80 km/h. The train is therefore going 20 km/h faster. If we refer instead

W. Grassi, *The Challenges of Time*,
https://doi.org/10.1007/978-3-030-94372-1_5

to the train, the car is slower than the train by 20 km/h, so we reach it and then we have the impression that it moves away from us with such speed in the opposite direction to the motion of the train. The passenger of the car who spontaneously takes this as his reference sees the train moving away from him at 20 km/h in the same direction of its motion. Let us suppose we arrive at the station of Grosseto and we stop next to another train. Who has not happened to not understand, for a few moments, which of the two trains has started? Of course, we look for some "fixed" reference through the opposite window and/or to perceive some jolt. If there were neither, we would have a hard time understanding how things are. When we are in a tunnel, the first thing that jumps out at us is the lights that "pass by" us. But if the tunnel were completely dark and the train perfectly silent and shake-free, would we really be able to tell if we were moving? We would see time passing on our watch, but we would not be able to tell if we had travelled any space.

In substance, the only way to be able to affirm if we are in motion or stopped is to specify with respect to what. This is, as we have already said, the principle of relativity formulated by Galileo, and it is valid for all those cases in which we refer to the so-called inertial reference systems, that is, those systems with respect to which if a body is initially stationary it remains at rest or if it moves with uniform rectilinear motion it remains in this state of motion without its velocity varying in modulus and/or direction. These reference systems are indistinguishable from each other and equivalent to describe physical phenomena.

In fact, up to now we have referred to systems (train or car) moving at constant speed in modulus and direction.

When we have mentioned possible "jolts" of the train that can make us notice its movement, we have in fact referred to moments in which the speed varies in time. In this case, as well as when the train starts or brakes to stop, we undergo accelerations (speed variations in time) and we clearly feel the effects of these.

So, if we observe what the train does between a station of departure and the next one of arrival, provided there are no stumbles during the journey and if the railway line does not change direction, we see its speed increases from zero to the value of regime, v, which it reaches after a certain time and maintains during the journey. From this moment, its motion is rectilinear and at constant speed. Before the arrival, the train starts to brake and its speed decreases up to zero. Both in the departure and in the arrival phase, the train is subject to speed variations. In the first case, it increases in time and the train is said to accelerate; in the second case, it decreases in time and the train decelerates. In both cases, there is an acceleration (i.e. we repeat, a variation of speed with time) that we are able to perceive even when we are sitting in the train or even more if we are standing in a crowded bus.

Let us remember that the speed is the ratio between a space x covered and the time interval that is employed to cover it t. If the space is one metre and the time taken to cover it is one second, the speed is one metre per second (m/s). So, if the train goes at 100 km/h, it means that it travels 100 km in one hour or 100.000 m in 3600 s (27.8 m/s). Starting from standstill, it will increase its speed from zero to 100 km/h. If in the first minute the speed becomes 10 km/h (2.8 m/s), it means that in one minute (60 s.) the speed variation is $\Delta v = v(60\ s) - v(0\ s) = 2.8{-}0 = 2.8$ m/s in a time interval $\Delta t = 60$ s. The average acceleration in this interval is the ratio between the velocity variation and the time in which this variation occurs $\Delta v/\Delta t = 2.8(\text{m/s})/60$ s $= 0.046$ m/s^2.

Let us exclude these phases for the moment and consider only those sections where the train is moving at a constant speed. Let us assume, again, that the available train fleet cannot exceed 150 km/h. This will be, therefore, the maximum attainable speed. If we indicate with O the point from which we start observing the movement of the various trains and with $x = 0$ and $t = 0$ its spatial coordinates along the railway line (always straight) and temporal coordinates ($t = 0$ indicates the value that our clock marks at the beginning of the observation) and we try to construct a graph that links the time that passes, t, to the space that is covered, x, we obtain a graph like the one in Fig. 5.1.

As we said the speed is the ratio between the spaces you cover divided by the time, it takes to cover it and so it is given by the slope of the straight lines that give the space as a function of time. The triangles in grey represent all possible train paths available (i.e. with a speed not exceeding 150 km/h). The segment OA_3 is relative to the fastest train model; in fact in one hour, it covers a distance of 150 km to the right on the railway line, for example, to Rome. Any other slower train, in the same time, can reach a shorter distance, segment OA'. A train stopped on a dead track can be denoted by segment OA_2, because after one hour it has not travelled a single metre. So, it has not moved, but of course time has passed anyway. Reserving the right to better specify the concept in the following, we always refer to the time of the observer at the point O.

What about the OA_1 segment? It is simply the segment covered by a train going, at the maximum allowed speed, to the left, i.e. in the opposite direction to OA_3. We can also put it differently. As things appear to us, the traveller who at the time $t = 0$ is in O will find himself (in future) in an hour in A_3 if the train is going at maximum speed towards Rome, in A' if it

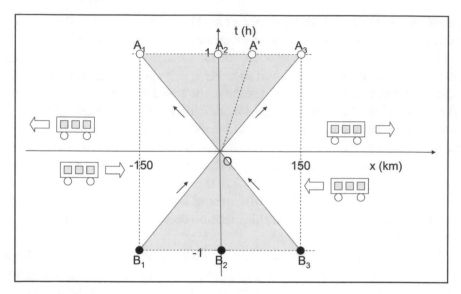

Fig. 5.1 Graphical representation of the distance travelled and the time taken by trains when there is a maximum speed that cannot be exceeded

is going slower and in A_1 if, with the faster train, it is going towards Genoa (opposite direction). Similar things can be said also for a traveller who is (in the past) one hour earlier ($t = -1$ h) closer to Rome or Genoa, or who got on a train that did not leave one hour ago. The area of the grey triangle corresponding to positive time, after $t = 0$, refers to what will happen in future, while the area corresponding to negative time (before $t = 0$) tells us about the past.

Sometimes, it is preferable to construct graphs that have homogeneous quantities, that is, expressed with the same units of measurement, on both coordinates. In this case, we can multiply the time by the maximum speed reachable by our trains that we indicate with V_0 and that, in our case, is 150 km/h. In this way, on the abscissas there remains the space x expressed in kilometres and on the ordinates $V_0 t$, expressed in these same units. The graph is modified as in Fig. 5.2. Doing so, the lines corresponding to the maximum speed of the trains coincide with the bisectors (45° inclined segments) of the axes of the graph. In fact, to a space crossed at the maximum speed, x^*, of 10 *km* we know that corresponds a time, t^*, of $10/150 = 1/15$ h, one fifteenth of an hour. The ratio between the ordinate $V_0 t^*$ (150 km $\times$ 1/15 km) and the space x^* (10 *km*) is equal to one. So, $V_0 t^*$ and x^* are numerically equal and are the sides of a square of which the segment OA* is the diagonal, which divides into two equal parts the right angle formed by those sides.

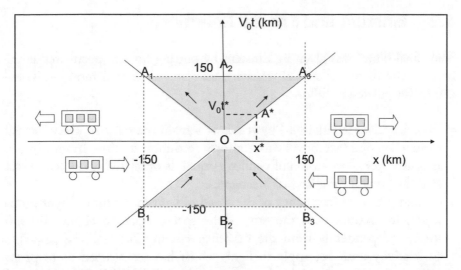

Fig. 5.2 Modification of the graph of Fig. 5.1, within ordinates the time multiplied for the maximum speed

By limiting the speed of trains, we have added to the principle of relativity of motion the impossibility to reach areas on the railway line in any time. For example, we know that starting now from the location O we cannot reach a place that is two hundred kilometers away in one hour. Theoretically, nothing prevents us from thinking this, but the reality of things docs, even if it is enough to wait for the railways to build faster lines and trains. That is, we just have to hope until this possibility is realized.

Now, however, we must take an important conceptual step that is counter-intuitive to our everyday experience. It, in fact, tells us that it is always imaginable to be able to increase the speed of an object. Technological progress has accustomed us to think this way, but is it always true? Let us try to understand it.

Up to now, we have understood that the absolute motion does not exist and that if there are some limitations to the speed with which we can move: we succeed however to reach any place but not always in a prefixed time. Path and time, in the case of a limit to speed, are not, therefore, independent. The question is: does nature place a limit on the speed with which we can move or with which we can send a signal? It seems so; nothing can move at a speed greater than the speed of light in a vacuum, commonly denoted by c.

We have seen in Chap. 3 which problems have raised the discoveries of Maxwell on the electromagnetic waves and how these had induced Lorentz to reach the conclusion that the values of the lengths had to depend on the reciprocal motion of the reference systems.

5.2   Einstein and Special Relativity

The "final blow" was given by Einstein by putting his two postulates at the basis of the theory of special relativity, referring to inertial motions. These can be formulated as follows.

- First postulate (relativity): Physical laws remain the same in every inertial system, i.e. one that moves with uniform rectilinear motion. In some ways, it constitutes an extension of Galileo's principle of relativity. Galileo in fact said the same for the laws of mechanics.
- Second postulate (constancy of the speed of light in vacuum): Light propagates in vacuum with the same finite speed, equal to about 300,000 km/s, independently from the reference system. The Galilean principle of relativity does not apply to light, which has accustomed us to properly compose speeds. Light always moves in vacuum with the same speed independently from the speed of the source that emits it.

We will accept them precisely as postulates, trying to see what consequences they bring to our conception of time. When I was a Boy Scout, we used to go on night hikes at camp in small groups armed with flashlights. In addition to light, we used them to communicate at a distance with a rudimentary Morse code. It was the easiest way, we did not have walkie-talkies, to stay in touch. We did not know it then, but we communicated in the fastest way: at the speed of light. By the way, in air it is essentially the same as in vacuum. We were sure we were doing it instantaneously. Later, after having studied physics in high school, we learned that the light from our torch covered the distance, for example, three hundred metres, to reach the companions of the other group, in a well-determined interval of time. So, the information we sent reached them with a certain delay. But what was this delay? Light goes at about three hundred thousand kilometres per second, which is three hundred million metres per second. Our signal arrived at its destination after a time given by the ratio between three hundred metres and three hundred million metres per second and therefore of one millionth of a second. Such a delay certainly did not give us any concern, and it was already difficult to perceive the passing of a second. Even if we had been three hundred kilometres away and could still see the light signal, the delay would have increased to one thousandth of a second, let alone one thousandth of a second.

But a bug is beginning to appear. What if the distances or times we are used to and refer to are no longer the same? We begin to remember that we only see the sunrise eight minutes after it has risen above the horizon. If we see the sunrise starts at eight o'clock on our clock, we can say that, if we could see the phenomenon instantaneously and then the light would propagate at infinite speed, the clock would mark seven fifty-two, with an advance $\Delta t = d/c$ compared to eight.

The sun is the closest star to us, and some estimates, of which we are not interested in giving more details here, assume that the diameter of the observable universe is of the order of more than 90 billion light years. The light year indicates the distance light travels in one year. If there are 31,540,000 s in a year and light travels a space of about 300.000 km every second, you do the math. These are distances we cannot even imagine. In addition, the universe is expanding and this brings some complications that we do not take into consideration, simplistically. What conclusion does all this lead us to?

When we look at the starry sky on a clear night, we feel small and lost, but as naive earthlings we think we are seeing a photograph of the "current" state of the visible universe. According to what we have said so far, we are not in the presence of a single photograph, but it is as if we were seeing frames of a film at the same time, one behind the other, shot at different times. The sun is the nearest star in both space and time, and the farthest stars are still in space and time. What we think we see at the same time, in our own time, is what happened to the object a short or long time before. Precisely because they transmit information to us with the same speed, that of light, the more distant they are in space the more distant they are in time. Space and time are thus intrinsically linked. When we see the light of the stack turns on three hundred kilometres away, it is what happened to the stack a millisecond before. If we send a signal to a space shuttle that travels, let us suppose, with uniform rectilinear motion in the universe, it receives it with a delay related to its distance from us and we cannot do anything about it. In the first case, we see a thing (light switching on) happened in our past by another individual standing still on the earth and therefore sympathetic to our same reference system. The space shuttle receives the signal we send it, in our future, while we have sent it in its past. The fact that there is an upper limit to the speed at which light information is transmitted means that these temporal differences are ineradicable. Or perhaps, it all depends on the fact that we refer to an absolute time just as we referred to an absolute space before Galileo.

Earlier, we posited in an essentially axiomatic way that the maximum speed attainable in nature was that of light, c, in vacuum and that this is a constant.

But Einstein says more, and that is that a "ray of light" always moves with velocity c, in vacuum, whether it is emitted by a body at rest or in motion. As Einstein wrote, "*We shall take this conjecture (the content of which will be called in the following "principle of relativity") as a postulate, and in addition to this we shall introduce the postulate with this only apparently incompatible, that light in empty space always propagates with a determined velocity V* [which we call *c* according to what has become the common usage], *independent of the state of motion of the emitting bodies"*. In this way, we are going to upset what we said about the rules for evaluating the velocity in relative motions between inertial systems. Perhaps, we will end up coming to the conclusion that that way of reasoning, considered by us so intuitive and logical, has been a sort of ball and chain of the mind that, before Einstein, prevented us from looking further. Since speed is the relation between space and time, it occurs to us, perhaps by rough reasoning, that when we approach the speed of light something happens to both space and time, such that their relation remains independent of the reference system we adopt. Therefore, time and space are no longer two independent quantities, since their ratio remains constant. We ask ourselves if this happens only when the speed approaches *c* and far from it everything goes as before, or it happens all the time and we are not able to appreciate it only because we are used to infinitely lower speeds? Do our senses deceive us?

Now we are confused. If we want to console ourselves a little, we can read the biography of Einstein written by Walter Isaacson [1]. In this biography, it is said that Einstein was also confused; he was stuck for more than a year on the contradiction that existed between the postulate of relativity of the motions and the one that sustained that the speed of light always had the same value whatever the inertial system referred to. He came to confess to his friend Besso that he was on the point of giving up. While he was talking to Besso, he had an intuition which he explained as follows: *"An analysis of the concept of time was my solution. Time cannot be defined in an absolute way and there is an inseparable relationship between time and the speed of signals"*.

In 1916, Einstein wrote a book to explain these points of view to the layman and we will make use of the example he gave. He also referred to the train and examined the point of view of an observer standing on the platform, whom we will call Jacopo, and of a woman travelling on the train, whom we will call Letizia.

The first one is standing at the newspaper kiosk (indicated with the point *M* in Fig. 5.3), and suddenly, he sees two lightnings falling, one on the right at a point *A* of the quay and the other on the left at a point *B*, equidistant from *M*, where he is. His clock reads the same time for both and so he knows that the two lightning strikes fell simultaneously.

In that same instant, marked by his watch, at rest with respect to the platform, Jacopo realizes that the seat on which Letizia is sitting placed at a point

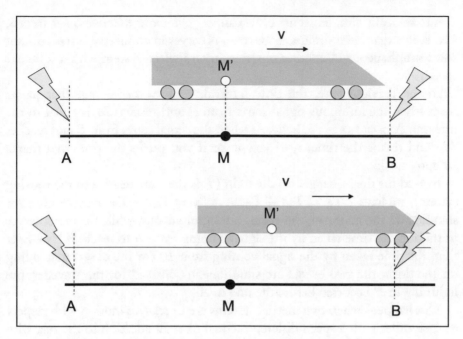

Fig. 5.3 Top: the observer on the platform sees the two lightning strikes simultaneously. Bottom: for the observer in motion, the two lightning strikes are not simultaneous

M' in the train coincides (it has the same coordinate x) with the point M on the platform, and it is just in front of the kiosk. But the train moves from left to right with a speed v. Letizia is moving from left to right with the same speed with respect to Jacopo, advancing towards B and moving away from A. According to the relativity of motion, the same as Galileo and Newton, Jacopo is moving with the same speed with respect to her, but in the opposite direction: from right to left. Since she is moving towards point B, Letizia will see first the lightning in this point, or more precisely the signal coming from it, and after the one in A. In fact, the lightning signal coming from B takes advantage of the fact that the train is moving towards it while it is moving away from A from which the signal reaches her later. Letizia's clock, on the system in motion, tells her, therefore, that the two lightning strikes are not simultaneous events.

In conclusion, for the observer standing still on the platform the two lightning strikes occur simultaneously, but not so for another observer who is on a reference system in motion with respect to the first. As we know, it is not possible to say who is stationary and who is in motion. The only thing we can say is that there is a relative velocity between the two observers. So, how is it possible to say what is the "right time", and how can we say that any two events are really simultaneous? How can we talk about *absolute time*? In fact, two events that seem simultaneous to an observer in an inertial system do not seem simultaneous to another observer in an inertial system with a different speed.

In the previous case, the time interval, $\Delta t_A = t_M - t_A$, that passes from when the luminous signal starts from A and reaches M is equal to the interval, $\Delta t_B = t_M - t_B$, between when the signal starts from B and reaches M. And this is the time, t, of Jacopo or if you prefer the time that marks Jacopo's clock.

Instead for the passenger on the train (T is the time relative to the moving system), we have $\Delta T_A = T_{M'} - T_A > \Delta T_B = T_{M'} - T_B$. For the observer standing on the platform, the events are simultaneous, while for the observer in motion the time taken by the signal coming from A to reach M' is longer than the time taken by the signal coming from B. For the observer standing on the dock, the two events are simultaneous; instead for the traveller, the lightning in B occurred before the one in A.

This happens when two inertial systems are in relative motion with respect to each other with a speed difference equal to v. In addition to the relativity of the motion of these systems, the concept of the relativity of simultaneity is affirmed. It is the idea of absolute time that enters in crisis.

To give an example that is perhaps more understandable, let us try to imagine that instead of two lightning bolts, two gunshots are fired at points A and B. The sound propagates at a speed of about three hundred metres per second, very slow compared to light; in fact, first you see the lightning and then you hear the thunder. Jacopo hears the two shots at the same time, and they probably sound like one shot. Letizia, who is travelling at a speed of one hundred kilometres per hour (twenty-eight metres per second), will first hear the shot in B and then the one in A. Let us do the math. The acoustic signal will proceed from B towards Letizia with a speed of three hundred and twenty-eight metres per second, while the one coming from A with two hundred and seventy-two metres per second. Between the two signals, there is, therefore, a difference in speed of fifty-six metres per second. It is logical that she perceives the sound coming from B first.

To decide who comes first in an athletic race of one hundred metres, we use a starting line and a finishing line, both fixed with respect to the earth, and we evaluate with very precise clocks the time interval, relative to the same reference system, that the athletes take to cover this distance. In order to have a significant time, we synchronize the chronometers; i.e. we make sure, by placing them next to each other, that the second hand is in the same position on their dials at the same time. We could also place the clocks one at the start and the other at the finish and emit a ray of light from the first at the moment of departure. In this way, we can establish, with practically no error for our purposes, the initial instant of the race (time zero). The second clock will take the time of the arrival of the athletes. If we indicate with d the distance finish line, start what is the error we make on the evaluation of the time of the race. If with t_P and t_t we indicate, respectively, the time in which the signal is emitted from the starting point and the arrival at the finish line and with c the speed of light, we have:

$$t_t - t_P = \frac{d}{c} = \frac{100}{300.000 \cdot 1000} = \frac{1}{3.000.000}(s)$$

As we have said, this error is practically inessential, since we expect the time to be evaluated to be a few seconds and we can safely neglect a third of a millionth of a second. In principle, however, synchronization as we commonly understand it is not perfect; it would be if the two clocks were exactly at the same point, or if we adopted some criterion to define such synchronization without having to bring the clocks closer together. The greater the distance between the points, the greater the error we make. It is necessary to formulate a criterion to define the simultaneity between the events and to define, on this basis, the concept of synchronization.

In order to clarify, we continue by drawing on an article [2] written by Einstein himself in 1905 on the electrodynamics of moving bodies. He focuses on the problem and then proposes a solution.

First of all, he points out how the concepts of time and simultaneity are deeply interconnected and tries to provide a first possible solution: "*We must keep in mind that all our assertions in which time plays a role are always assertions about simultaneous events. When, for example, I say: "That train arrives here at 7 o'clock" this means: "The setting of the small hand of my watch to 7 o'clock and the arrival of the train are simultaneous events. It might seem that all the difficulties concerning the definition of "time" could be overcome if I were to substitute for "time" the expression "position of the small hand of my watch"*".

He then stresses the need to refine the evaluation method in order to take into account events occurring in different positions, just as happens when a light signal is sent from the starting zone to the finishing line: "*Such a definition is sufficient when it is a matter of defining a time independently of the position in which the clock is located; but the definition is no longer sufficient when it is a matter of temporally linking series of events occurring in different places, or - which is equivalent - temporally evaluating events occurring in places far from the clock. We could also be satisfied with the temporal evaluation of events by means of an observer who is at the same time as the clock at the origin of the coordinates, and who associates the corresponding position of the hands of the clock with every light signal that reaches him through the empty space, and that bears witness to the event to be evaluated. However, such a co-ordination brings with it the drawback of not being independent from the point of view of the observer looking at the clock, as we know from experience".*

He examines the case of two observers in different positions and, as he says, arrives at: "*a much more practical determination by the following consideration".* He establishes a common time for the two observers by calling into question the speed of light and by giving a definition of synchronism based on one of his usual mental experiments, as follows: "*If at point A in space there is a clock, an observer at A can temporally evaluate events in the immediate vicinity of A by observing the positions of the clock hands simultaneous with these events. If also at point B of the space there is a clock - we add, "a clock with exactly the same properties as the one at A" - then a temporal evaluation of the events in the immediate surroundings of B by an observer at B is also possible. It is not possible, however, without further deliberation, to compare temporally an event in A with an event in B; so far we have defined only a "time of A" and a "time of B", but we have not defined any "time" for A and B altogether. The latter time can be defined only when we assume by definition that the "time" that light takes to go from A to B is equal to the "time" that it takes to go from B to A.*

That is, a ray of light starts at "time of A", t_A, from A towards B, is at "time of B", t_B, reflected towards A and returns to A at "time of A" t'_A. The two clocks by definition walk synchronously when

$$t_B - t_A = t'_A - t_B ".$$

This relation is not valid if one of the two points moves with a certain velocity v, with respect to the other as we have learned from the mental experiment of Jacopo and Letizia. In fact, in the case of a moving observer and a fixed observer, so not both at rest with respect to the same reference system, the signals coming from the points where the lightning had fallen were seen as simultaneous by the observer attached to the platform, but no longer by the person travelling on the train.

Continuing with Einstein, "*Let us assume that this definition of synchronism is possible in a contradiction-free way, that therefore the conditions apply:*

1. *When the clock in B walks synchronously with the clock in A, the clock in A walks synchronously with the clock in B.*
2. *When the clock in A walks synchronously with both the clock in B and the clock in C, the clocks in B and C walk in mutually synchronous ways.*

We have thus determined with the help of certain (thought) physical experiences what is to be understood by rest clocks walking synchronously and being in separate places, and with this we have evidently obtained a definition of "simultaneous" and "time". The "time" of an event is the simultaneous indication with the event of a clock at rest which is in the position of the event, walking synchronously with a given clock at rest, and that is, for all the determinations of time made with the clock itself".

We assume according to experience that the quantity [AB is the distance between A and B in the reference at rest]:

$$\frac{2AB}{t'_A - t_A} = c$$

is a universal constant (the speed of light in empty space) c. It is essential that we have defined time by means of clocks at rest in the system at rest; we call the time now defined, because of this association with the system at rest "the time of the system at rest"".

We have understood that in two inertial systems moving in parallel with respect to each other, the events are not synchronized. Let us call time into play, and with another experiment, perhaps abused but effective, let us try to understand something more about it. Let us imagine that Jacopo, who is a shrewd experimenter, is using a device consisting of a laser with which he sends a light impulse to a mirror, placed above him and distant l, in a direction orthogonal to the dock. The laser is also equipped with a detector that receives the ray reflected by the mirror. Jacopo measures the time interval that the pulse takes to go to the mirror and back to the detector by travelling a round trip distance of $2l$. This interval, Δt (t is the time evaluated by Jacopo in the system at rest), is equal to the distance $2l$ divided by the speed of light c: $\Delta t = 2l/c$. Letizia, who is looking out of the window and is passing through the same position as Jacopo at the moment the impulse is sent, moving to the right, sees the equipment moving away to the left and, therefore, a path of the impulse that is no longer orthogonal to the platform.

Let us stop for a moment to consider the situation in detail. Let us think that instead of the light ray a small ball is thrown with a vertical speed, w, which remains constant in modulus and inverts its sign coming back after bouncing in the mirror. We can also think of following a photon that, however, in this case moves with speed w instead of c. Figure 5.4 shows how Jacopo sees things and how Letizia sees them. For the first one, the ball assumes the positions 0, 1, 2, 3, shown in (a), at the times $t_0 = 0$, t_1 ($\Delta t_1 = t_1 - t_0 = t_1$), t_2 (Δt_2) and $t(\Delta t)$ in which it has covered the distance l. Differently Letizia, who sees Jacopo moving away with velocity v

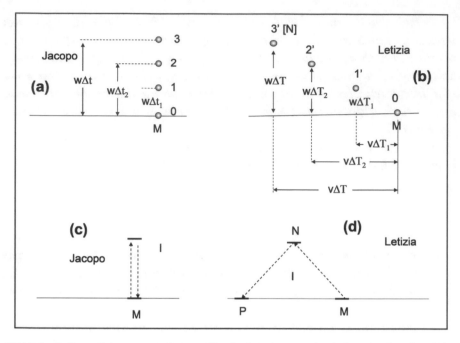

Fig. 5.4 Ball motion as seen from a fixed observer, on the left, **a, c.** On the right, the same phenomenon seen from a moving observer, **b, d**

(in modulus) towards the left, sees them assuming the positions 0, 1', 2', 3', shown in (b). While the ball moves on a vertical stretch with velocity w in an interval ΔT (measured with her clock), it moves to the left of a stretch $v\Delta T$. The stretches of space according to which the ball moves for Jacopo and Letizia are shown in (c) and (d).

However, this is a Newtonian point of view. According to this point of view, while Jacopo would see the ball travelling along the vertical tracts with speed w, Letizia would see the tracts MN and NP travelling with a speed given by the composition of v and w. On the other hand, from Einstein's point of view, if the ball represents the light signal, both strokes are travelled with the same velocity c. In this case, things go as in Fig. 5.5 which shows the two paths of light as seen by both observers. According to Letizia, the light starts from point M and arrives in N after a time $\Delta T = T_N - T_M$ of the moving[1] system and finally in P, after the same time since the train proceeds in uniform rectilinear motion. The paths MN and NP, equal to each other, are the hypotenuses of right triangles of catheters of length l and $v\Delta T$ and are equal to the square root of the sum of the squares of l and $v\Delta T$. The

---

[1] It is in fact she who sees Jacopo's equipment moving.

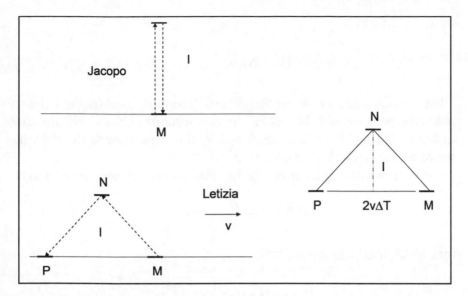

Fig. 5.5 Paths as seen by the stationary observer (top) and the moving observer (bottom and right)

traveller, therefore Letizia, estimates an interval of time, $2\Delta T$, necessary to complete the path MN + NP. This interval is given by this length divided by the speed of light, which, as we now know, is still equal to c even if referred to an inertial system in motion. The first conclusion we can draw is that the elapsed time $2\Delta T$, for the observer in motion (Letizia), who sees the phenomenon starts and ends at different points M and P, is greater than that elapsed, $2\Delta t$, for the observer at rest (Jacopo), who in turn sees the phenomenon starts and ends at the same point.

Before continuing, we must pay attention to what we are doing. If we refer to the ground, Jacopo, on the dock, is at rest (stationary with respect to it) and Letizia is moving to the right.

If, instead, we refer to the train on which Letizia travels, it is the platform that moves to the left, while she is stationary. From this perspective, the time, t, measured by Jacopo is that of a moving clock while the time seen by Letizia, T, is relative to a stopped clock.

The time referred to those who see the phenomenon begins and ends at the same point is called proper time.

Therefore, based on the situation we were observing for the time intervals, and we can reason as follows. Jacopo sees to cover the length l twice, once in outwards and the other in return with the speed c. The total distance is equal to two times the length l ($2l$) and is completed in a time $2\Delta t$, measured on his watch, where:

$$\Delta t = \frac{l}{c}$$

For Letizia, who, let us not forget, sees Jacopo in motion, the light ray makes the path we said. In each of the two sections MN and NP, the time needed is ΔT (with Letizia's time) and in the same interval the train has moved on the track of a section $v2\Delta T$.

For each of them, again remembering Pythagoras' theorem, we find that:

$$(c\Delta T)^2 = l^2 + (v\Delta T)^2$$

from which it is easily derived that:

$$(\Delta T)^2 \left(1 - \frac{v^2}{c^2}\right) = \frac{l^2}{c^2}$$

If we remember the value of the time interval, Δt, we see that:

$$\Delta T = \frac{\Delta t}{\sqrt{\left(1 - \frac{v^2}{c^2}\right)}}$$

Finally, we obtain a link between the time of the moving observer (in this case Jacopo with the interval Δt) and the time of the stationary observer (in this case Letizia with the interval ΔT):

$$\Delta T = \gamma \Delta t$$

$$\gamma = \frac{1}{\sqrt{\left(1 - \frac{v^2}{c^2}\right)}} > 1$$

The link between these two intervals is γ (>1) and is the Lorentz factor already mentioned. This means that the clock of the moving observer goes faster than that of the stationary observer, as in Fig. 5.6.

Or, if you prefer the clock of the stationary observer moves more slowly.

If Jacopo picks up a pen from the ground that he has dropped, Letizia sees him do it much more slowly (over a longer interval of time) than she would.

If Jacopo, moving relative to Letizia, saw the experiment lasts a certain period t, Letizia saw it last longer. If an event for Jacopo lasts one hour, for

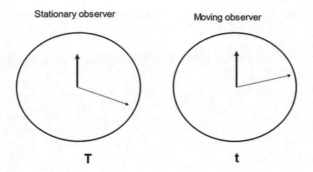

Fig. 5.6 Time of a stopped clock (left) and one in motion (right)

Letizia it lasts, for example, an hour and a half. Or: if Jacopo's clock marks the passing of an hour, Letizia's clock marks an hour and a half. We could also say that while those who are in motion age by one hour, those who are stationary age by an hour and a half.

We did an experiment that recalled a bit the procedure of Michelson and Morley and that made explicit a relation between the time measured by a mobile observer and that measured by a fixed one, highlighting, as Lorentz had already said, the dilation of time.

From the previous relations we understand why, at the speed v with which we normally deal with, the two time intervals appear substantially equal. In fact, the square of the ratio between this speed and the speed of light is very small and essentially not appreciable until the speeds of objects and/or signals are not close enough to c. To get an idea, we see that it is equal to 1.005 when the ratio v/c is equal to 0.1. In other words the variation of the time interval of the system in motion with respect to the proper time (or better to the proper time interval) is five thousandths when the velocity v is worth 30,000 km/s, which means 30,000 multiplied by 3600 (=108,000,000) kilometres per hour. More in detail Fig. 5.7 shows the graph of the values of γ as a function of the ratio of the velocities.

Bertrand Russell [3] beautifully clarifies the meaning of proper time with these words: "*The universal cosmic time that was taken for granted is no longer admissible. For each body there is a definite order of time concerning the events that take place in its vicinity; we may call it the 'proper time' of that body. Our experience is governed by the "proper time" referring to our body. Since all of us remain more or less stationary on the earth, the proper times of different human beings agree and can be united in a single time, called earth time. But this is only the time referable to the terrestrial bodies of great dimensions*". In fact, if, for example, we refer to the elementary particles the speech becomes more articulated and we will say something about it later.

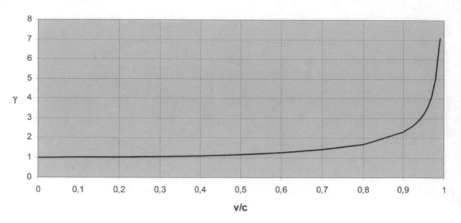

Fig. 5.7 Factor γ as a function of the speed ratio

What conclusions do we come to after all this? What causes us to consider different times? The answer is: the relative motion of one observer with respect to the other. Let's try to put on the train the equipment that Jacopo uses on the ground. Letizia sees the impulse moving up and down vertically, and therefore, it starts from a point and falls on the same point. The time interval that Letizia measures is the same t that Jacopo measured and vice versa, Jacopo sees things as Letizia saw them before, only the direction of the relative speed changes, and consequently, he measures the same T that Letizia measured before. While before it was Jacopo's time measured by Letizia that flowed more slowly than his, now it is the opposite.

Eddington (1882–1944) in his book "Space Time and Gravitation" (1920), referring to an aviator flying at a speed close to the speed of light, says: "*If we (we on earth) carefully observe the aviator, we will deduce that he is unusually slow in his movements and everything that happens in the aircraft that moves with him will appear to us similarly slowed down, as if time had forgotten to flow. His cigarette lasts twice as long as ours….*".

Let us complete the reasoning referring again to our characters Jacopo and Letizia, and let them make an experiment, the first one on the platform and the second one on the train. Let us consider again two points A and B on the platform, integral with it, and C and D on the train[2] (integral with the train). Let the distance AB be the same as CD, AB = CD = d_0, at rest (this indicates the subscript 0 to the distance d), i.e. measured in the reference system of the platform. When the train is at rest, with M coincident with M', the clocks in A and B are synchronized as are those in C and D. With

---

[2] We could take as the length to be measured the wheelbase between the wheels, or bogies, of the wagon.

the same notations as before, we can say that $t_B - t_A = t'_A - t_B = \text{AB}/c = d_0/c$ and $T_D - T_C = T'_C - T_D = \text{CD}/c = d_0/c$. The clocks in A and B are synchronized with each other $\left(t'_A - t_A = 2d_0/c\right)$ and so are the clocks in C and D with each other $\left(T'_C - T_C = 2\text{CD}/c = 2\text{AB}/c = 2d_0/c\right)$. To synchronize all four of them, it will be enough to take $t_A = T_C$, being the positions of A and C coincident.

If, on the other hand, the train is moving relative to the platform by a speed v to the right, can we say the same?

Letizia, with respect to which points C and D remain fixed, still sees things as before. To ascertain the value of the distance between points C and D, Letizia measures it, while the train is moving, with the same metre with which she had measured it when the train was stationary. This measurement gives her the same result; if before it was ten metres she finds the same value, or rather her metre is ten times the distance between C and D. At the moment, we can only say that the measuring metre in motion is still equal to one tenth of the distance between the points in motion. We already know what happens to clocks, but let us talk about it again. Jacopo sees the light C starts at the instant t_C and it proceeds towards D with a speed, with respect to the train, $c - v$, because the point D is moving away with a speed v towards the right. Then, it is instantaneously reflected back into D (which has moved into D') and returns into C (which is moving into C') with a velocity $c + v$, because C is "coming towards it" with the same velocity v. The superscript (') refers to the points that have moved from their initial positions C and D.

To visualize what's going on, let us look at Fig. 5.8.

At the top is drawn the stationary wagon. At the bottom, the wagon is moving with speed v and always in uniform rectilinear motion. It is represented in three different moments. In the first, drawn with a continuous line, the signal starts from C and goes towards D, which is moving in the meantime. The second, with a dotted line, represents the instant in which the light ray reaches D, which has moved to position D' with respect to the track, and is reflected back. The third, dotted, is the instant in which the reflected ray reaches C, which, however, in the meantime has moved to C'. In the same figure, the velocities are indicated.

Consequently, $t'_{D'} - t_C = d/(c - v)$ and $t'_{C'} - t'_{D'} = d/(c + v)$. We have written d and not d_0, even if to Letizia the distance continues to result the same. In fact, we have become a bit suspicious and the fact that it still turns out to be equal to ten times the measuring tape does not lead us to conclude that it is the same for everyone, but only that the ratio between distance and measuring tape is the same. Moreover, we verify, if there were still a need, that the clocks are not synchronized even though they were synchronized

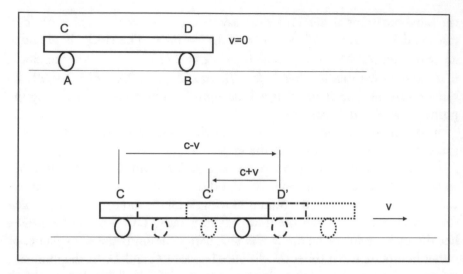

Fig. 5.8 Carriage stopped at top; carriage moving at bottom in three different instants: continuous lines at the instant when the light ray starts from *C*, dashed lines when the ray arrives in *D* (moved to *D′*) and is reflected, dotted lines when the reflected ray arrives in *C* (in the meantime moved to *C′*)

both among themselves and with those on the platform when the train was stopped. This is another confirmation of the fact that simultaneity depends on the inertial reference system.

We can also say that:

$$t'_{C'} - t_C = d\left(\frac{1}{c-v} + \frac{1}{c+v}\right) = \frac{d}{c}\frac{2c^2}{c^2 - v^2} = \frac{2d}{c}\frac{1}{1 - \frac{v^2}{c^2}} = \frac{2d}{c}\gamma^2$$

For Letizia, things continue to go as before. A ray of light starting from *C* starts at a time T_C and returns, once reflected, to *D* at a time $T'_{C'}$.[3] Therefore, it can be written:

$$T'_{C'} - T_C = \frac{2d_0}{c}$$

---

[3] Let us repeat again that Letizia sees *C* and *D* as stationary, so T'_D marks the instant of return (this is the sense of the apex) of the light ray at point *D*.

But we know that when the train is in motion Letizia's time (T), from Jacopo's point of view, dilates with respect to his (t); that is, it passes more slowly:

$$\Delta t = \Delta T \cdot \gamma$$

from which we get

$$\gamma = \frac{t'_{C'} - t_C}{(T'_{C'} - T_C)} = \frac{d}{d_0} \cdot \gamma^2$$

$$d = d_0 \cdot \frac{1}{\gamma}$$

We were right to be suspicious, while time expands on the system in motion, seen by the stationary observer, space, seen by the same observer, contracts, since γ is greater than 1. If when the train is stationary, both see the same distance d_0, and when it moves, Letizia still measures d_0 and Jacopo sees that the measuring tape is contained ten times in this distance. At the same time, he sees d_0 contracting into d and the metre has also suffered the same fate. This is the same conclusion Lorentz had reached. So, the stationary observer sees that time passes more slowly than his, but also that the lengths in the direction of motion are shortened. Ultimately, for Letizia the distance between the points of the wagon remains the same as when it was stationary, and for Jacopo it becomes shorter. We repeat that all this was obtained by imposing what the second postulate of special relativity states, that is, that the speed of light always has the same value.

But we, large terrestrial bodies, as Russell says, do not notice this, even though we are used to seeing things of different sizes depending on the distance from which we look at them. When my dog runs around in the meadows, I see him change size as he moves away. I know very well that he is about fifty centimetres tall and seeing him become smaller in perspective than a cat does not surprise me. He alters his "apparent" size, as all things do if I get closer or farther away. I say "apparent" because I am convinced that his real dimensions are those when he is standing near me, just as I say that my friend's watch is synchronized with mine because I see his hands mark the same hour and minutes as mine when he is near me.

If we were small and moving at speeds close to the speed of light, the alterations in time and space would appear much more evident. A classic example for everyone can be made by referring to a particle called muon. The muon is an unstable particle (identical to the electron, but heavier) that forms due to the interaction between cosmic rays and the upper layers of the atmosphere and has a decay time of about 2.2 millionths of a second, giving rise to an electron and a pair of neutrinos. Attention: The decay time we are talking about here is the one of a reference integral with the muon, with respect to which it is therefore stationary. But let us not go into details about the decay of the particles, which are not of interest here. What interests us instead is to know that they go at a speed almost equal to that of light (99.9% of it: $v = 0.999c$). With such a high speed and such a decay time, which is the life time of the particle, a simple calculation of the product of speed and time, as we were used to do for a uniform rectilinear motion, tells us that the muon should travel in the atmosphere a little more than six hundred and fifty metres. Yet, we find them on the earth in great quantities, after having traversed the whole atmosphere, how is this? What we earthlings see. The muon is very fast compared to us who are "still", and therefore, we see its time passes much more slowly, so much so that we judge that its life is about twenty-five times greater and reaches fifty-five millionths of a second. Those who wish to verify this can do the math with the formulas we have given. With the speed at which it moves, the muon can travel sixteen kilometres and cross the atmosphere. To go back to the previous examples, we are standing still on the platform and the muon is the passenger on the train coming towards us and moving with the speed we have said: its time is dilated. The length of the path it takes in the atmosphere "stationary with respect to us" can only be the product of the particle's speed and the time interval we measure. The muon, on the train, is essentially stationary with respect to it. For it, its existence still lasts two millionths of a second and it sees us, on the platform, moving towards it with a velocity v, which has the same value with which we see it coming towards us, but in the opposite direction. Even the atmosphere is not stationary with respect to him. He therefore sees the path to be taken contracted twenty-five times more and thus reduced to the above-mentioned six hundred and fifty metres. So, we are all agreed.

To avoid having generated confusion, I want to make, once more, a clarification. Speaking of contraction of space, we must keep in mind that it is the dimension in the direction of motion, parallel to the velocity v, that shrinks, those orthogonal to it are preserved and are not, therefore, subject to the same phenomenon: the wagon becomes shorter, but its height and width remain constant. I said before that by the effect of perspective I see my dog smaller when he is far from me. But in this case, all the dimensions change while maintaining the mutual proportions. In the context of what we have been discussing, things are a little different. Think of a basset hound you know those dogs with the big ears, long body and short legs. If he could run very fast, but really very fast, you would see him stay the same height but take on more harmonious proportions by shortening.

At the conclusion of the paragraph, in the following reading, we will say more about how space and time are transformed.

Mathematical digression—The dilation of time and the contraction of space.

Let us keep on a plane configuration with ordinates given by time t and space x for the fixed observer (Jacopo) and T and x' for the mobile observer (Letizia), in which the motion occurs parallel to the x-axis in its positive direction and the origins of the two systems ($x = x' = 0$) coincide for $t = T = 0$. Lorentz gives the following formulas to pass from one system to the other.

$$x' = \gamma(x - v \cdot t)$$
$$T = \gamma\left(t - \frac{v}{c^2}x\right)$$
$$\gamma = \frac{1}{\sqrt{1 - \left(\frac{v}{c}\right)^2}} > 1$$

Let us look at these formulas a little bit by referring to the previous discussion.

If we look at it from the point of view of Galilean relativity ($v \ll c$), which is referred to in Fig. 5.9, we obtain the classical conclusions about two reference systems in relative motion, as can be the ones solidified by Jacopo and Letizia.

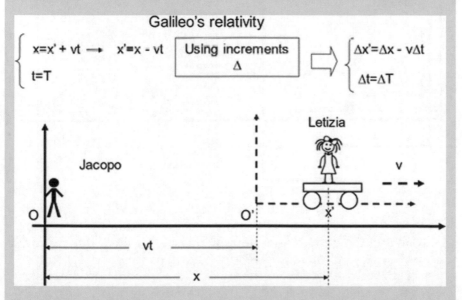

Fig. 5.9 Letizia's motion with respect to Jacopo according to Galileo's point of view

If Jacopo and Letizia are in the positions of Fig. 5.9, we would say that the sum of the coordinate x' of Letizia with respect to the reference system in motion, with velocity v to the right, plus the distance travelled from its origin O' with respect to that of the fixed system O, OO', is equal to the distance of Letizia from the origin, O, of the fixed system, from which we obtain:

$$x = OO' + x' = vt + x'$$

We have, however, said that from the point of view of special relativity, Jacopo sees time dilate for Letizia and space contract in the direction of motion. For the sake of convenience, let us also remember that, while Jacopo sees Letizia moving away towards the right with velocity v, Letizia sees Jacopo moving away in the opposite direction that is with velocity $-v$. Let us refer to Fig. 5.10, which is more detailed.

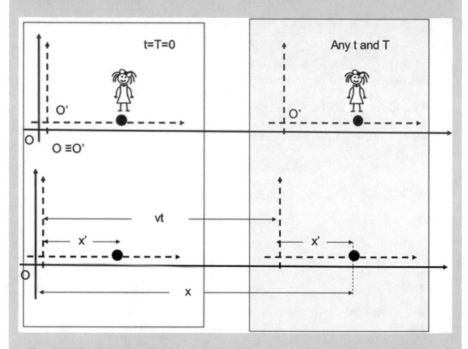

Fig. 5.10 Representation of Letizia's motion with respect to Jacopo. The continuous line indicates the stationary reference system and the dashed line the moving one. At the initial instant ($t = T = 0$, left panel), the spatial origins coincide ($O \equiv O'$) and Letizia's distance from them is x'. At a generic instant (right panel), Letizia's distance from the moving system is x' and x ($= vt + x'$) from the fixed system

Jacopo sees the segment $OO' = vt$ and x with this value of length, but he sees the segment of length x' contracting, multiplied, consequently by a factor of $1/\gamma$. We have:

$$x = \frac{x'}{\gamma} + vt$$

Or, even:

$$x' = \gamma(x - vt)$$

If instead we consider the mobile reference, x' is seen with its length (on this reference), the segment x moves with velocity $-v$ and, for this reason, it is contracted with the origin O that moves away from O', evaluating the time on the mobile reference

$$x' + vT = \frac{x}{\gamma}$$

Substituting the expression of x', we get:

$$\gamma(x - vt) + vT = \frac{x}{\gamma}$$

$$x\left(\gamma - \frac{1}{\gamma}\right) = v(\gamma t - T) \rightarrow x\frac{(\gamma^2 - 1)}{\gamma} = v(\gamma t - T)$$

And since

$$\frac{(\gamma^2 - 1)}{\gamma} = \sqrt{1 - \left(\frac{v}{c}\right)^2}\left[\frac{1}{1 - \left(\frac{v}{c}\right)^2} - 1\right] = \frac{\left(\frac{v}{c}\right)^2}{\sqrt{1 - \left(\frac{v}{c}\right)^2}} = \gamma\left(\frac{v}{c}\right)^2$$

You get the relationship we were looking for

$$T = \gamma\left(t - \frac{vx}{c^2}\right)$$

5.3   A Bit of Minkowski Geometry

For most of us, geometry is Euclidean geometry, essentially flat. Yet we live on a roughly spherical planet and the difficulties of representing its geography on a map are many. Think about how big Greenland appears on a map. It looks bigger than Canada, and yet, it is extended "only" a little more than two million square kilometres, while the surface of Canada is about ten million. No wonder we are representing parts of a sphere on a sheet of paper.

Coordinates on earth are given as a function of longitude and latitude. Let us call into question the parallels, which are circles with an ever-smaller radius going from the equator to the poles, and the meridians, circular arcs connecting one pole to another and perpendicular point by point to the parallels. Let us make two brief reflections. The parallels meet at the poles. Yet, Euclid tells us that two parallel lines never meet. Let us construct a triangle formed by a section of the equator and the parts of the meridians that start from the extremes of this section and join at the north pole. It looks just like a triangle, but on a curved surface. The parallels meet the equator perpendicularly and, therefore, form with it angles of ninety degrees. When they then meet at the pole, they form a larger or smaller angle depending on which ones we have chosen. On a sheet of paper, the sum of the internal angles of a triangle is one hundred and eighty degrees, but not those of a "spherical triangle".

We will stop here because this is not intended to be a lesson in geometry, but only to show how there is not only one.

In the following, we will mention a geometry that not only considers spatial coordinates, which we normally indicate with variables x, y, z in a three-dimensional space, the only kind of space we can visualize, but also a fourth coordinate: the temporal one. As we did for the trains, we will use only one spatial coordinate, x, the same one that the railway line travels, in order to simplify the discourse. Moreover, we will continue to refer to bodies that move with uniform rectilinear motion.

But let us see what we are talking about. We could say that it is the geometry of space–time, a new concept that somehow jumps out of the things we have discussed so far. It was born by Herman Minkowski (1864–1909), a professor of mathematics at the Zurich Polytechnic, a Jew of Russian origin who was much appreciated by Einstein, but who, not reciprocating this sentiment, initially called him a slacker who did not care at all about mathematics. On the other hand, the beginnings of Einstein's academic history are peculiar: he failed a physics course, and his first doctoral thesis was rejected by the Zurich Polytechnic.

Years later, Minkowski, surprised and amazed by the work of his former student, converted the concepts expressed by Einstein into mathematics. In Isaacson's book, already quoted, we read: "*Einstein, who was not yet in love with mathematics, dismissed Minkowski's work as "superfluous erudition" and concealed: "Since mathematicians have taken over the theory of relativity, I myself do not understand it anymore". But in fact, he came to admire Minkowski's work and devoted a paragraph to it in his 1916 popular book on relativity*".

To describe this approach, we start by recalling what we said before about trains not exceeding the speed V_0 of 150 km/h (Fig. 5.2). Now, we have a speed that is never possible to exceed in nature: it is the speed of light. We should no longer speak of trains, but of signals, elementary particles, etc. We will still use the language of us "terrestrials" referring, as far as possible, to everyday life. Figure 5.11 becomes a figure in which the x-axis is still that of space and the ordinates, instead of reporting the product of the maximum speed reachable by trains for the time, considered as maximum speed that of light. The first big difference is that, while for trains we could always hope that their maximum speed would increase thanks to the progress of technology, for light we do not have this hope. The bisectors of the four quadrants, or rather their inclinations, this time, correspond to the speed c, and the grey areas are called "cones of light".

The space axis (x-axis) corresponds to time zero, i.e. to the present. Below this is the past, to which negative times correspond, while above this is the future with positive times. If we got sick yesterday morning, at the time corresponding to B_2, now we are still in bed, point O, and we will stay there until after tomorrow morning (A_2) when we will get up, the segment that represents our state throughout this period is B_2A_2. The more shrewd may notice that, at least to go to the bathroom, we have to cover some space.

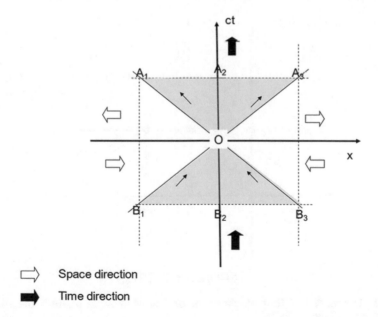

Space direction

Time direction

Fig. 5.11 Minkowski plane. The same as in Fig. 5.2 with c instead of V_0 on the ordinates

They are right, and then, without going into details, let us say that we are also forbidden to get up because we have a strong labyrinthitis. However, it is clear that whatever movement we make or whatever event or signal we refer to, our life takes place within the grey triangles. The reference system, which we may call system 1, S_1, we are talking about is "stationary" in the sense that, while we are in bed, we do not make any movements. We have a clock by which we can take the times (ours) of what is happening at some distance from us. If we maintained the mentality of Galileo and Newton, who advocated absolute time, we would see the simultaneity of events as follows. Let us remember that anyway in everyday life we perceive them this way. For example, at the instant t_1, just after drinking a glass of water, which we will call event B, we see through the window the light of a streetlamp on the street (we are always trivializing events that, in order to be appreciated, should happen at infinitely greater distances), event L (the lighting of the streetlamp), and simultaneously the lights of the stadium turn on, event F. Where are these events placed in Minkowski's plane, maintaining the conception of absolute time? (Fig. 5.12)

These are events (switching on the streetlight and the headlights) stationary in space that occur at different positions. They lie, therefore, on a line parallel to the x-axis, which would be the line of simultaneity for reference S_1.

But this does not obey the criterion of simultaneity established by Einstein. Let us begin to see how things are for a reference in which the observer does

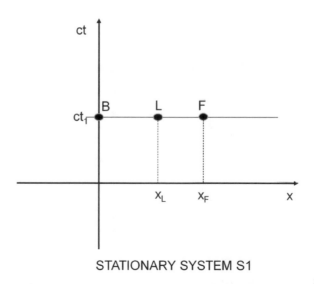

STATIONARY SYSTEM S1

Fig. 5.12 Events B, L, F that would occur on the Minkowski plane if the speed of the light signal were infinite

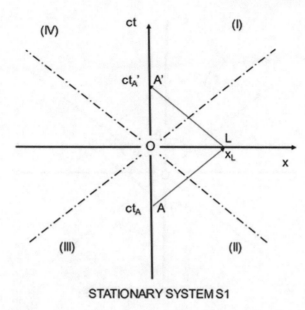

STATIONARY SYSTEM S1

Fig. 5.13 Happening of the event L if the B would happen at the instant $t = 0$

not move in space, but only in time, or as we might say his universe line[4] coincides with the axis of time. Suppose we have sent a light impulse in the past (point A) towards the lamp post in x_L. It will proceed with the speed of light (parallel to the bisector of the first quadrant, the quadrants of the plane are indicated by Roman numerals in Fig. 5.13) towards the lamp post and reach it in a time interval $\Delta t = t_O - t_A$ (t_A is negative). In our representation, we have assumed $t_O = 0$ (the present corresponds to $t = 0$) and, being in the past, the time of A is in the zone of negative values. However, the length of the interval is equal to $t_O - t_A$. This interval is equal to the length of the segment AL divided by the speed of light: $\Delta t = AL/c$. Meanwhile, the observer, initially at position A, moves $c\Delta t$ along the time axis. The ray, reflected in L, returns towards $x = 0$ with velocity c (parallel to the bisector of the second and fourth quadrants: it goes back in space but continues in

---

[4] The universe line is the set of points traversed by the subject in the Minkowski plane. An observer who does not move has a universe line coincident with the time coordinate. If a reader, fed up with the subject, decides to kill the observer (stationary in $x = 0$), at a certain instant ($t = 0$), this instantaneous event reduces to a point. At the moment he kills it, t is equal to zero, present, and the point coincides with the origin. If we continue to observe Minkowski's plane, we see the time of the event going further and further down in negative values (in the past).

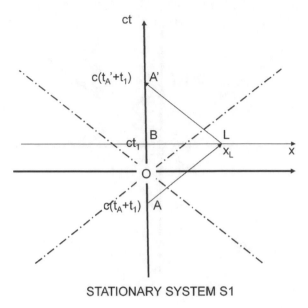

STATIONARY SYSTEM S1

Fig. 5.14 Happening of the event L if the B happens at the instant $t = t_1$

time) and arrives at point A' after a time $\Delta t' = A'L/c = AL/c$. With simple geometrical considerations,[5] we see that $AO = OA' = c\Delta t$.

The path of light is $AL + LA' = 2AL$ and $c = 2AL/(t'_A - t_A)$, which is precisely the condition of simultaneity. In conclusion, we can deduce that, for this reference, the line of simultaneity coincides with the x-axis.

To also include the fact that we started after drinking the glass of water at time t_1, we can translate the previous graph by one term ct_1 and make the same considerations as in Fig. 5.14.

If instead we look for the line of simultaneity for a system moving with velocity v, we start by drawing its universe line, that is, the line representing its space–time evolution in the Minkowski plane. We know that this line forms with the x-axis an angle greater than forty-five degrees ($v < c$). Let us indicate with T the time referred to the moving system (Fig. 5.15). At the beginning of the observation, the origins of the two systems, still and in motion, are considered coincident. With the passing of time T, also t passes in the same direction but with different durations as well as the position of the moving subject changes with respect to the fixed reference.

---

[5] The segment AL forms with the x-axis an angle ALO, with vertex at L, of 45°, and the same happens for the angle A′LO. The two right triangles ALO and A′LO have the cathetus OL in common. It is easily seen that $A'L = AL = OL/sen(45°)$.

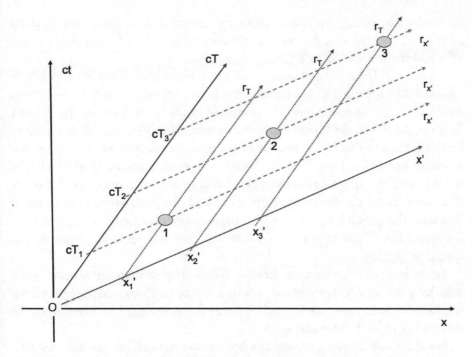

Fig. 5.15 Representation of a mobile reference system. The axes of time and space change at cT and x'. At time $T = 0$, the axis of space intersects it at $x' = 0$. As time T passes, it moves parallel to itself and meets it at successive points: cT_1, cT_2, etc. Similarly, it happens for that of time cT

In the stationary system, the slope of the bisector line is, as said, the ratio $ct/x = 1$. Since the speed of light does not depend on the inertial system to which we refer, it must also be $cT/x' = 1$ and this happens only if the axis x' is symmetrical to cT with respect to the bisector. It is emphasized that the new metric that is introduced in the space–time plane is no longer the Euclidean one to which we are accustomed.

Let us remember that on a stationary reference point, with the passage of time, the line representing space moves parallel to itself, intersecting that of time at different values. In its turn, the one of time moves to the right intersecting the one of space in successive positions. In any case in this reference, the straight lines of time and space remain perpendicular.

If the observer is in motion, the new reference system has coordinates cT and x', and if we consider the case in which initially also this system has coordinates $T = 0$ and $x' = 0$, things go as in Fig. 5.15. With the passage of the time T, of the system in motion, the straight line of the space x' moves progressively, like the ones we have indicated with r_x, and intersects the one of the time (always at $x' = 0$) in successive times, T_1, T_2, T_3. Similarly, proceeding in space, the line of time, r_T, meets that of space (for $T = 0$) at points x_1', x_2', x_3'. So, the point 1 has coordinates x_1', cT_1 the point 2: x_2', cT_2 and the point 3: x_3', cT_3.

An event F (Fig. 5.16) at time t_F and at position x_F in the fixed system is represented with the point of coordinates (with respect to the fixed reference) x_F and ct_F. To represent it on the system in motion, we draw for this point a line, $r_{x'}$, parallel to the line of simultaneity and a line, r_T, parallel to the "time line" cT for $x' = 0$, of the system in motion. As mentioned, the space line x' moves upward as time T of the observer in motion passes. With the latter, we intersect the line of simultaneity of the system in motion at the point F of coordinate x_F'. By an analogous method, i.e. intersecting the universe line with the parallel, $r_{x'}$, to the simultaneity line, we find the elapsed time distance cT_F. These (x_F' and cT_F) are the coordinates of the event in the system in motion.

From here, we understand how a space–time system in motion with velocity v is transformed with respect to a "stationary" one. More precisely, a system with a universe line that admits only temporal displacements into one that moves in both time and space.

For the more curious, or perhaps braver and masochistic we can demonstrate all of this in the box below, which others can skip over without inconvenience.

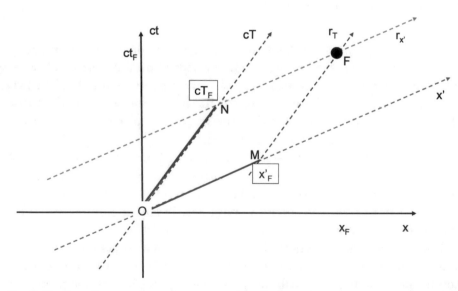

Fig. 5.16 Coordinates of an F-event in the Minkowski plane, for a fixed and a moving reference

Mathematical digression—Line of simultaneity for a system in motion and other considerations

Let us follow the same procedure used before for the stationary system, and let us refer to Fig. 5.17. With point A, we indicate our position in motion at a certain instant T_A (point A is still with respect to the moving system). After an interval of time ΔT, this point will pass through O, after which it overcomes it, reaching a point A' at the instant T'_A. Suppose that on the line of simultaneity, of which we do not yet know the position, is distributed a large number of mirrors capable of reflecting a light impulse in all directions. We do not see them because we are at night and it is possible to locate them only at the moment in which they reflect the light. From position A, we send a light impulse that will proceed with the speed of light (AP parallel to the bisector of the first and third quadrant). We are interested only in the reflected impulse that will reach us at the point A' where $\Delta T' = \Delta T$ that is after a time of the system in motion equal to that which the ray takes to reach the mirror. This will reach us again with the speed of light following the direction of the bisector of the second and fourth quadrant. The point of intersection (the two lines intersect perpendicularly, both having a slope of 45°), P, of the line followed by the impulse sent by A at time T_A and of the one that reaches A' at time $T'_{A'}$ lies on the line of simultaneity whose coordinate we indicate, for now, with y.

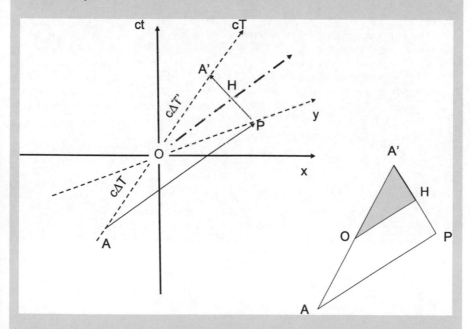

Fig. 5.17 Graphical construction of the reference system in motion

Since the point A passes through O, the latter also belongs to the above line. The triangle APA' is a right triangle, $2\Delta T = (AP + PA')/c$, and simple mathematical calculations confirm that $AP = PA' = c\Delta T$.

On the right, further down, are drawn, for greater clarity, the two triangles OHA' right in H, and APA', right in P. We know that $AA' = 2OA'$. The two triangles are similar so we also have $A'P = 2A'H$ and $A'H = HP$, and the bisector of the angle A'AP is still the bisector of the first quadrant. From this, we deduce that the coordinate tentatively denoted by y coincides with x'. We have taken for convenience the system that refers to the dock with origin at $x_0 = 0$ and $t_0 = 0$, and the x-axis indicates the place of the points where the clocks integral with it always give the time $t = t_0 = 0$. We could have started from another origin O' with $x_{O'}$, $t_{O'}$. On the new x-axis, the clocks would all give the time $t_{O'}$. The same happens for the x'-axis of the system in motion, where all clocks indicate the same time $T = T_O$.

One thing we can immediately see is that the lines cT and x' have relations that can be expressed as a function of the coordinates of the fixed system. The slope of cT is given by c/v, and ct is related on it to x by the equality:

$$ct = \frac{c}{v}x = \frac{x}{\beta}$$

$$\beta = \frac{v}{c}$$

The line x' is symmetrical to this one with respect to the bisector, which has a unit slope (tangent of 45°). If we denote by α the angle formed between this line and the bisector, the angle formed by x' with the x-axis is $(45 - \alpha)$ and its tangent tang $(45 - \alpha)$ will be the slope of the line with respect to the x-axis. The angle formed by the universe line, cT, with the x-axis is obviously $45 + \alpha$ whose tangent is equal to c/v. With some trigonometric calculation,[6] it can be found for points on x':

$$ct = tg(45 - \alpha) \cdot x$$

$$ct = \frac{v}{c}x - \beta x$$

Let us recall for convenience the Lorentz transformations, which will be very useful in the continuation of this discussion.

---

[6] $tga = 1 \quad a = 45°$

$tg(a + \alpha) = \frac{c}{v}$

$ct = tg(a + \alpha) \cdot x = \frac{c}{v}x$

$tg(a + \alpha) = \frac{tga + tg\alpha}{1 - tga \cdot tg\alpha} = \frac{1 + tg\alpha}{1 - tg\alpha} = \frac{c}{v} \rightarrow tg\alpha = \frac{c-v}{c+v}$

$tg(a - \alpha) = \frac{tga - tg\alpha}{1 + tga \cdot tg\alpha} = \frac{1 - tg\alpha}{1 + tg\alpha} = \frac{c+v-c+v}{c+v+c-v} = \frac{v}{c}$

$ct = tg(a - \alpha) \cdot x = \frac{v}{c}x$

We will use these relations transforming them appropriately to use them easily in the space–time plane.

$$T = \gamma\left(t - \frac{v}{c^2}x\right)$$

$$x' = \gamma(x - v \cdot t)$$

$$\gamma = \frac{1}{\sqrt{1 - \left(\frac{v}{c}\right)^2}} > 1$$

We get

$$cT = \gamma\left(ct - \frac{v}{c}x\right) = \gamma(ct - \beta x)$$

$$x' = \gamma\left(x - \frac{v}{c}ct\right) = \gamma(x - \beta ct)$$

from which we get:

$$ct = \frac{cT}{\gamma} + \beta x$$

$$ct = \frac{x}{\beta} - \frac{x'}{\gamma\beta}$$

For $T = 0$ and $x' = 0$, we arrive at the previous relations.
Let us reconsider Fig. 5.16 (Fig. 5.18).

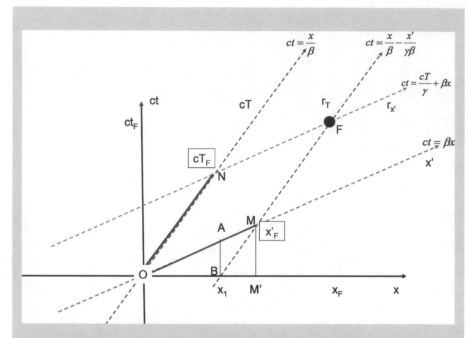

Fig. 5.18 Like Fig. 5.16 in which, however, the mathematical bonds between spaces and times are highlighted

The lines r_T intersect the x-axis at the points where $t = 0$: $x = x'/\gamma$. Similarly, the x' lines meet the time axis, ct, for $x = 0$ at the points $ct = cT/\gamma$.

If the stationary observer sees the event in x_F and at a time t_F, the moving observer sees it in $x_{F'}$ and T_F given by:

$$cT_F = \gamma(ct_F - \beta x_F)$$
$$x'_F = \gamma(x_F - \beta ct_F)$$

After all this math, since our health has improved, we get out of bed and go to lunch and, just to give a patina of scientificity to an everyday situation, and let us see how we can bring it back to the Minkowski plane. We get out of bed and go, step by step, to the kitchen. We are very presumptuous to think that we can represent all this in the space–time plane. In fact, given the speeds and the distances involved, going from the bed to the kitchen, staying there for a while to eat and going back to bed is a set of events that would remain indistinguishable. But by greatly enlarging the scale of the spaces and, as usual, considering that accelerations are absolutely negligible (we always move with uniform rectilinear motion and the bed and the kitchen table are separated by a long straight corridor), our actions can be represented as in Fig. 5.19 (not to scale).

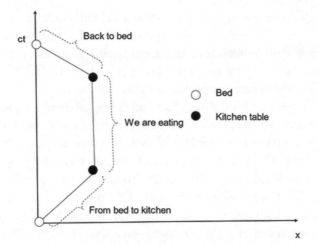

Fig. 5.19 Minkowski plane description of the actions we perform

When we go from the bed to the kitchen, both space and time change. While we eat space does not change, but time passes. Going to bed again, we reverse the path in space and return to where we were, but, unfortunately, time continues to pass.

If this were not so, we might think that time goes back to a time before we got up. We would have been satiated before we got out of bed. Yet, we got up precisely in order to be satiated. For example, we got up at one o'clock to go eat, but we go back, travelling backwards through space, and crawl back into bed at noon with a full belly.

But we got up precisely to fill our stomach, that is, the cause for which we got up was to feed ourselves. The effect produced by that cause was to feed ourselves. The two events of getting up and feeding are therefore linked by a principle of causation. To be able to go back to bed full an hour before we got up violates this principle that, instead, at present we consider inviolable. Linked events (I get up and I get full) must obey this principle.

Beware of the term concatenated events, by which we meant causally connected events. They are such: I get up and I feed myself but not I get up and John feeds himself. The fact that I get up has nothing to do with the fact that John feeds himself unless I have to get up because it is I who prepare his lunch. Einstein does not exclude the possibility of temporal reversal of events as long as this reversal does not change anything in the functioning of the universe. For example, I cannot go back in time and be born before my parents.

Many would say that a way like this of approaching the problem of causality is, if we want to be benevolent, simplistic. We do not blame them, indeed, but at the moment we try to get around this consideration by saying that we want to refer only to our experience of the macroscopic world of which we have daily perception. If I have health problems because of gall-stones, I go to the hospital and have an operation. It is difficult to think of going back in time before the operation without the stones.

Let us look again at how this aspect can be represented in the space–time plane. Let us use an example. We are near the gate of our house waiting for a truck to deliver a piece of furniture. We see the truck coming, and we decide to open the gate. The first event (event 1) is the decision to open the gate. With the key in hand, we set off, at an instant t_1, to open it. The second event (event 2) is the opening of the gate. Our universe line is a straight line (we always move with uniform rectilinear motion) with a slope greater than 45°, because we move at a slower speed than that of light. The space–time path is the segment 1–2 in Fig. 5.20. If instead of having to move we could open the gate, which has a special detector, with a laser signal and we wanted

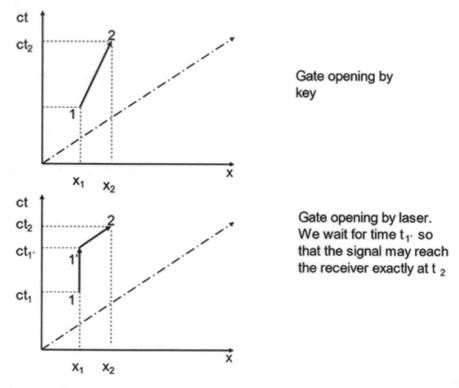

Gate opening by key

Gate opening by laser. We wait for time $t_{1'}$ so that the signal may reach the receiver exactly at t_2

Fig. 5.20 Two ways of opening the gate. At the top we walk, starting at time t_1 from a certain position x_1, to open the gate with the key (event 2). At the bottom, we can open with the remote control, so we can wait a while (from t_1 to $t_{1'}$) before giving the impulse for the gate to open at time t_2

to open the gate exactly at the instant t_2 that is at the exact instant when the truck reaches the gate, we can wait a little more time (1 to 1′) and send the signal so that it reaches the detector at the desired instant (2). Segment 1–2 is parallel to the bisector being sent at the speed of light. If we wait too little, the gate opens sooner, and if we wait too long, it opens later and the truck has to wait before entering.

Wanting to use a language just a little more rigorous, we say that two events 1 and 2 in space–time can interact along some space–time paths. In the first case (opening with the keys), the universe line of 1 passes through 2 with velocity v. Since v is less than c, this is a possible path. In the other case, the universe line of 1 is first parallel to the ct-axis (at fixed space only time is passing) and then changes to a line parallel to the bisector line. If $t_{1'}$ were too small, event 2 would not be reachable at time t_2, just as if $t_{1'}$ were too large.

Remember that event 2 corresponds to the opening of the gate at position x_2 at exactly time t_2.

Two such events, one the cause of the other, are said to be "causally related".

Think about how much more dramatic it can be if a signal of failure from a space probe three hundred million kilometres away from earth arrives in Houston (e.g. the average earth–Mars distance is two hundred and twenty-five million kilometres). Whoever receives the signal knows that it was sent about seventeen minutes earlier and that it takes just as long for a response signal to reach the probe. It will also take some time to figure out what the type of failure might be and the probe continues to move. Will they be able to intervene effectively or in the meantime will the failure have caused the destruction of the probe, undoing years of work? In this case, event 1 may be the arrival of the fault warning to Houston, 1′ the sending of the appropriate signal to repair it from Houston to the probe and 2 the arrival of the signal to the probe.

Before leaving the subject, it is good to note that the term "macroscopic" has peeped out again just as it did when we talked about the arrow of time. Our daily experience and the conceptions that have been established over centuries of the history of thought confirm the validity of the category of cause and effect.

The issue is extremely delicate and goes so far as to call into question determinism to the point of undermining free will. It is far from us to enter into these problems. We will limit ourselves to briefly recalling Heisenberg and his principle of indetermination.[7] He affirms that deterministic laws have value, but their applicability is put in crisis by the impossibility to know the state of a system with absolute precision. This impossibility does not depend on the characteristics of the instruments but is intrinsic to nature. While for us large earthlings, as Russel says, causality is an unavoidable pillar of the understanding of things and it works, at the size of the electron we run into the uncertainty principle that seems to make us renounce it.

It seems as if nature does not want to reveal all the cards and let us see them up to a certain point. Or maybe it is us who still have not understood what game we should play.

Let us take now on a different plane. So far, we have seen that in addition to the principle of the relativity of motion, which has been known for a long time, the principle of the relativity of time also applies because of the constancy and the independence from the reference system of the speed of light in vacuum. "Constancy" and "invariant" (a quantity that does not vary when certain conditions change) are words that physicists have always liked a lot. A constant entity is one that is conserved by passing from one situation to another. So, we speak, for example, of conservation of momentum and energy and physics is full of constants, from that of ideal gases to that of Planck. It comes to us to wonder if it exists in the plane of the space–time that in fact it is a space to four dimensions, and let us not forget it, something that is maintained constant beyond c.

Let us look again at the Lorentz equations written for the space–time plane:

$$cT = \gamma(ct - \beta x)$$
$$x' = \gamma(x - \beta ct)$$

The invariant quantity we seek would have to be such whether we refer to the coordinate system x, ct or the coordinate system x', cT. In other words, this quantity would represent something invariant in nature, in which space and time are inextricably linked regardless of the reference system.

Reasoning in the context of Euclidean geometry, which we are accustomed to, we would refer in the plane to a figure characterized by a property for which an algebraic relation between the coordinates is maintained constant. The first to come to mind is the circumference, which is the locus of points equidistant from the centre. The value of the radius is a combination of

---

[7] In two words, the principle says that it is not possible to know simultaneously the position and the velocity of an elementary particle with absolute precision. The more precisely we know one, the greater is the error we make in the knowledge of the other.

the two coordinates (square root of the sum of their squares) that remains constant. Let us try to follow this path and see what happens if we take a circumference with centre at point O and radius R. Applying the Pythagorean theorem, a pillar of Euclidean geometry, we have the relation: $R^2 = (ct)^2 + x^2$. At least for the stationary observer, we would have concluded, except for verifying that there are no incompatibilities for the rather particular plane we are dealing with.

Talking again about the displacements we make in going from the bed (point O) to the kitchen (point P or P') and referring to Fig. 5.21, can we think of making them according to the path OP or OP'? If you remember what we said about the shaded areas, cones of light, and those not, you will immediately say that OP is possible but OP' is not because it imposes to move with a speed greater than that of light. We cannot reach the point P' starting from O because it would mean to cover the space $x_{P'}$ in a time $t_{P'}$, with a speed that is impossible to reach.

But let us also look at what is implied by the position of point P'', which is also in a zone of the cone of light. If we get up from the bed in O and we arrive in the kitchen in P'', is everything OK? For space, there are no contraindications: we go to a position at a certain distance from O, $x_{P''}$. But what happens to our clock: it says one o'clock when we get up and five to one when we get to the kitchen. If we follow the OP path, instead it tells us that it is 1:00 when we get up and 1:05 when we get to the kitchen. Obviously, taking R as invariant leads us to contradictory conclusions and in contrast to the principle of causality.

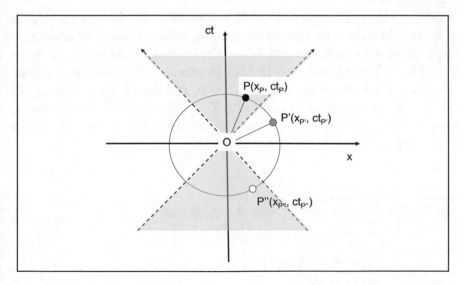

Fig. 5.21 Attempt to find the invariant

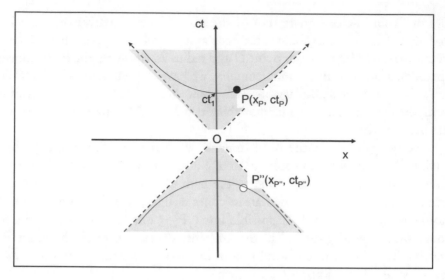

Fig. 5.22 Representation of the hyperbolic curve of the invariant quantity

We are therefore far from having found something physically significant. Someone will say: of course, you have reasoned according to the metrics of a different geometry. That is right, but habit is hard to die. In fact, an invariant quantity exists and can be found by changing the plus sign with the minus sign in the sum of squares of the radius formula.[8] This invariant, usually denoted by the letter s, is therefore: $s^2 = (ct)^2 - x^2 = (cT)^2 - x'^2$. The curves obtained are qualitatively drawn in Fig. 5.22. Mathematicians call them hyperbolas.

Let us go back to Jacopo and Letizia, one standing still (always with respect to the earth) on the quay and the other one travelling, sitting and reading her book. What does Letizia think? Given the interest of the book, she never moved from her seat and took an hour to complete the journey. The space she travelled, relative to the train, is $x'^* = 0$, and the time it took her is $T^* = 1$ h. The value of she gets is $s^{*2} = (cT^*)^2 - 0 = (cT^*)^2$. For Jacopo has completed a route of 100 km, $x^* = 100$ km. How much time, t^*, do you think has passed?

---

[8]
$$(cT)^2 - x'^2 = \gamma^2\left[(ct)^2 + (\beta x)^2 - 2ct\beta x - x^2 - (\beta ct)^2 + 2\beta ctx\right]$$
$$= \gamma^2\left[(ct)^2 + (\beta x)^2 - x^2 - (\beta ct)^2\right]$$
$$= \gamma^2\left[(ct)^2\left(1 - \beta^2\right) - x^2\left(1 - \beta^2\right)\right]$$
$$(cT)^2 - x'^2 = (ct)^2 - x^2 = s^2$$

Also for Jacopo, we have the same s^{*2}. The time t^* can, therefore, be obtained from the invariance of s in passing from the train system to the platform system. He writes, therefore:

$$\left(ct^*\right)^2 - x^{*2} = s^{*2} = (cT)^{*2}$$

Since he measures a train speed[9] v^* of one hundred kilometres per hour, he knows that $x^* = v^*t^*$. He then arrives at the result:

$$\left(ct^*\right)^2 - \left(v^*t^*\right)^2 = s^{*2} = \left(cT^*\right)^2$$

$$t^{*2} = \frac{s^{*2}}{c^2 - v^{*2}} = \frac{T^{*2}}{1 - \left(\frac{v^*}{c}\right)^2}$$

$$t^* = \gamma T^*$$

Letizia's trip for Jacopo took longer than the hour she had calculated.

Mathematical digression on the construction of the invariant curve

These curves are entirely included in the light cone areas. In fact, the minimum value that s^2 can assume is zero. In correspondence,

$$v^2 = \left(\frac{x}{t}\right)^2 = c^2$$

$$\frac{x^2}{(ct)^2} = 1$$

The hyperbola degenerates in the straight lines limiting the light cones and for $x = 0$ also $t = 0$. For values different from zero, the intersection between the curve and the time axis ($x = 0$) is $s^2 = (ct)^2$. Thus, the curve passing through the point $P_1 \equiv (x = 0, ct_1)$ has value $s_1^2 = (ct_1)^2$. The values of x and ct that correspond to the same value of $s = s_1$ are, therefore, given by $(ct)^2 - x^2 = s_1^2$. The same is true for the coordinates of a moving reference: x' and cT.

Logically, there exists an infinity of s^2 curves according to the value that s assumes and that coincides with the intersection with the axis ct. We can take as an example the value $s = 1$.[10] The universe line of a moving system that

---

[9] Once again, we do not consider the phases of acceleration at the start and deceleration at the finish line.

[10] To put it better, $s = \pm 1$, with $s = 1$ in future (positive times), and $s = -1$ in the past (negative times).

as we know has equation $ct = x/\beta$ intersects the curve s^2 in point P (or better event) in which we have:

$$s^2 = 1 = (ct)^2 - x^2 = \left(\frac{x}{\beta}\right)^2 - x^2$$

And then at the abscissa:

$$x^2 = \frac{1}{\frac{1}{\beta^2} - 1} = \frac{\beta^2}{1 - \beta^2} = (\beta\gamma)^2$$

$$x = \pm\beta\gamma$$

$$(ct)^2 = 1 + x^2 = 1 + (\beta\gamma)^2$$

Consequently, the universe lines intersecting the curve as a function of the velocity v with which the system moves are infinite.

The line, $r_{x'}$, parallel to x' and therefore with the same slope (same angular coefficient) as it passes through the point of abscissa $x = \beta\gamma$ has the following equation (Fig. 5.23):

$$ct = \frac{cT}{\gamma} + \beta x = \frac{cT}{\gamma} + \beta\gamma$$

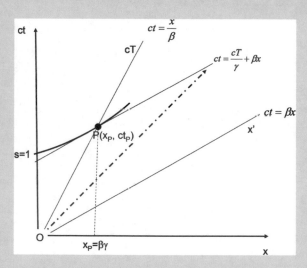

Fig. 5.23 Construction of the invariant curve

The curve $s = 1$ is expressed by:

$$(ct)^2 = 1 + x^2$$

And its slope from:

$$\frac{d(ct)^2}{dx} = 2ct\frac{d(ct)}{dx} = 2x$$
$$\frac{d(ct)}{dx} = \frac{x}{ct} = \frac{v}{c} = \beta$$

So, slopes of the line $r_{x'}$ and of the curve $s = 1$ coincide.

5.4 Does Time Depend Only on the Observer?

So far, we have talked about observatories on inertial systems and we have gone on trains moving at constant speed. But what if we wanted to become children again and go on a merry-go-round, perhaps to accompany our young child or grandchild? You will tell me we certainly do not think of Einstein; I will confess: neither do I. But let me take this as a cue to introduce the discourse on acceleration, moving from special to general relativity with regard to time. It is the same acceleration that is at the basis of the force that we would feel pushing us towards the periphery of the rotating base if the carousel rotated with a greater speed. It is the same as the force that the clothes are subjected to in the washing machine's centrifuge. We are no longer discussing inertial systems.

The first step was taken by Einstein with the so-called equivalence principle, according to which it is not possible to distinguish between an accelerated system and a stationary one subject to the force of gravity.

I had the luck to make for my work some flights on the plane that European Space Agency (ESA) calls zero-g and uses to simulate weightlessness. For this purpose, the aircraft, starting from level flight, parallel to the ground, climbs to an angle of 47° with the horizon. Figure 5.24 shows the trajectory that follows.

During the climb (phase AB in the figure), we feel crushed with force to the floor and it is said that we are subjected to an acceleration about double that of gravity that we feel on earth. The feeling is that the floor of the plane is coming towards us and pressing down on our feet. In fact, we cannot blame the gravity exerted by the earth, since we are at a distance from the ground

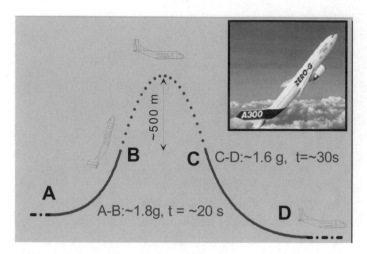

Fig. 5.24 Trajectory followed by the aircraft during a parabola

which varies insufficiently to make us feel the effects of the change in the so-called gravitational attraction. However, we feel like we are standing still but on a planet with a gravity acceleration twice the one we are used to. Anyone who remembers a bit of physics knows that we are subject, thanks to the pilot's manoeuvres, to an acceleration that produces this sensation. Then, in the CD section of the trajectory, a ballistic trajectory, the engines practically cease their thrusting action, compensating only for air friction and are turned back on at point *D*. Meanwhile, we follow the same parabola traversed by a projectile fired from a cannon. This is the period simplistically called zero-gravity, zero-g. Now, we "float", and we do not feel any forces pushing us towards any part of the nacelle. If we are not anchored with straps to the floor we float, it is not uncommon for some stomach to do the same. Figure 5.25 gives an idea of this phase. In the foreground, you can see two of my fellow adventurers, next to our experimental equipment. I am "sitting" behind Roberto Manetti, from behind, the indispensable creator of the device, while Prof. Di Marco (in those times a research assistant I introduced to microgravity research) is testing his stomach in an upside-down position. The rest of the picture shows those who follow the experiences, those who take advantage of the evolutions and those who, lying down, try to keep their stomachs happy.

It is as if we were standing still in space, far from any planet or more precisely from anybody capable of attracting us.

We have not certainly pretended to explain rigorously the principle of Einstein, but only to try to put together what it implies and that is the indistinguishability between an accelerated reference system and one firm

Fig. 5.25 A moment of the microgravity phase

and subjected to external force of gravity. Let us keep in mind that in all the circumstances of which we have spoken we are firm in comparison with the airplane and that when we say to feel subject or not to an acceleration of gravity we think about us that statically feel such an effect.

But we are back to the merry-go-round. When it starts to turn, we get off and leave the child on the little horse. We are the fixed observer, and the child is the mobile one, even if he does not know it and anyway he could not care less. We measure the radius and the circumference of the wheel of the stationary carousel. As we know from elementary geometry, the circumference is as long as the product of the radius by two times pi: $C = 2\pi R$.

What happens if the wheel begins to turn at six revolutions per minute which is 1 tenth of a revolution per second. It travels through an angle of thirty-six degrees per second or even one-tenth of a 2π (360° expressed in radians) which is worth about 0.63 radians. A point taken on the circumference of a wheel of radius five metres travels, during its circular motion, a space equal to the product of the radius itself for the angle. In one second, the length of the arc crossed is equal to $0.63 \times 5 = 3.15$ m. Its speed v is therefore a little more than three meters per second. Of course, we do not notice anything.

What would we see if everything moved with incomparably greater velocities? The radius of the wheel, keeping itself always orthogonal to the direction of the motion, remains the same that we have measured when stationary, but the circumference, whose points move according to the direction of the speed, contracts. And here comes into play the famous γ. In fact, the length of the circumference contracts by this factor, for those who are stationary with respect to the ground. Here, we are again arguing with Euclid; we have a circumference of radius R, but with a length less than that which geometry tells us. The force acting on the little horse of the merry-go-round, from which we have prudently taken away the child, is the greater, the greater the speed of rotation of the merry-go-round. In conclusion, the greater the speed, the greater the force and the greater the contraction of the length of the circumference.

On the basis of the principle of equivalence, we come to say that the stronger is the gravitational attraction and, therefore, the greater or closer is the mass of the attracting body, the more important is the contraction effect. Similarly, the time of the observer in motion dilates and we can make a statement analogous to the previous one about this dilation and the masses present.

At this point, a fundamental Einstein's "conjecture" intervenes and it has been experimentally confirmed many times. Let us see what it is about taking into account that on the horse, due to the rotation, a (centripetal) force is exerted towards the centre of the wheel that has to balance the centrifugal one. It is the same force that a hammer thrower must exert before launching or that you feel if you want to throw a stone with a slingshot. The horse, the hammer, the stone are subject to a force that could also be caused by a "stationary" body that attracts them because of its mass. According to Einstein, it is the space that changes and that is why the circumference changes its length. This modification now is due not only to the effect of the relative motion among reference systems, but also to bodies endowed with mass (or if you want of energy, on the base of the famous relation $E = mc^2$, on which we would not dwell). The situation, still referring to the carousel, is shown in a qualitative way in Fig. 5.26.

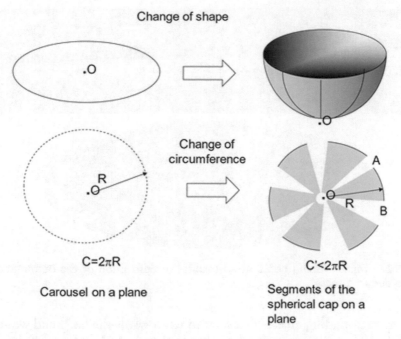

Change of shape

Change of circumference

$C=2\pi R$

Carousel on a plane

$C'<2\pi R$

Segments of the spherical cap on a plane

Fig. 5.26 Change of a circle, left, into a spherical cap, right, due to contraction of the circumferences

On the left is the stationary carousel disc. It has the same radius R for both observers. On the right, the carousel moves and the circumference contracts.[11] If the circle is transformed into the surface of a spherical cap (we have not chosen this shape at random) and we open this cap on a flat surface, the length of the circumference is transformed into the sum of the arcs of the circle that are obtained by "opening" each segment on the plane of the paper. The new length is obtained by summing the lengths of the various arcs AB, each still having radius R, which remains unchanged even for the observer in motion.

To explain the effect of a body on space, let us take a banal example hoping that Einstein and the specialists in the field will forgive us. Have you ever seen a fishing net, like those used along rivers or on the coast of the sea. If, in particular, you look at the net when you pull it up from the water and you have done good fishing, it is concave in the middle, accentuating its original shape. This shape has the advantage of conveying the catch towards the centre and preventing it from escaping. Something similar happens to space in the vicinity of a body, as shown in Fig. 5.27. And what is true for fish is also true

---

[11] The length of the circumference is reduced, but the distance between each of its points and the centre O remains the same. The centre is now on the ridge of the spherical cap.

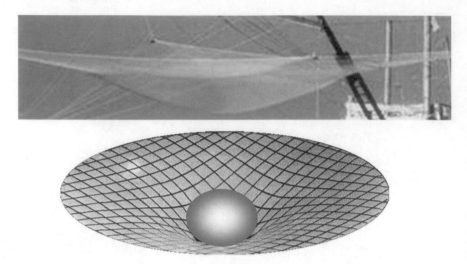

Fig. 5.27 Top: A fishing net. Bottom qualitative description of the deformation of space due to a mass

for any surrounding objects that tend to fall towards the body and would do so if they were not moving as the many satellites orbiting the earth do.

But if space changes, time must also pass more slowly in the presence of a gravitational field, just as it did for moving trains.

Let us return once again to the merry-go-round. We have seen that the speed of a point rotating on its disc is the smaller the distance from the centre. This also means that the contraction of the circumference it travels is smaller and the slowing of time greater. The same thing happens for the points closer to the centre of the body it attracts. For the benefit of those who live in the penthouse of a skyscraper, time passes more slowly for those in the basement. Imagine what would happen if we were able to appreciate such small differences in time, or if these differences were so large as to affect the length of human life.

Real estate agencies would be flooded with requests for basements, while no one would want penthouses.

After these facts had been demonstrated by various researchers, James Chin-Wen-Chou of the National Institute of Standards and Technology found, in a 2010 experiment published in the Journal Science, that the effects are detectable without moving away from the earth's surface.

Chou and co-workers placed two very precise atomic clocks 33 cm apart vertically. They thus showed that the lower clock marked time more slowly. The bad news is that for a human being living seventy-nine years, life time is extended by ninety billionths of a second. The positive news is a further demonstration of the validity of Einstein's theory about gravitational attraction or rather the modification of space due to the presence of masses.

The same researchers have, with a following experimentation, further demonstrated the equivalence between the presence of masses and forces, according to the principle of equivalence.

Let us change now a little perspective, and let us go back to the light rays we have widely used in the previous discussion. We have often spoken about them, but implicitly we have always intended to refer to their, for us, usual rectilinear path as it happens in vacuum or in air at uniform temperature.

But we know that under certain conditions, this property is lost. An example for all is the so-called mirages that occur in the desert but also on hot days on the road. In the second case, you can see in the distance the asphalt glistening as if there were water. More suggestive is the one that makes you see things that are there[12] but are beyond the horizon and not as close as they appear. These are all optical effects related to the distribution of air density, generally caused by differences in temperature, and consequently of the refractive index. Without going into details, it is sufficient to say that the composition of the air is no longer uniform and causes a curving of the path of light rays. It is as if the space near us had a different conformation from the uniform one we are used to. If similar situations occurred at night, we could see a light that we would otherwise not see because it was beyond the horizon or behind some obstacle. But in the desert at night, it is cold and there is no hope of that happening.

A brilliant physicist, Sir Arthur Eddington, had the idea that, without any intention of disrespecting him, we try to describe in a few simple words. If space is, as Einstein says, so strange that it behaves as we said around a body, it should be possible to see a luminous signal coming from a source even if hidden from us by the body itself. Something is similar to what described in Fig. 5.28. The light ray sent by the object has a curved trajectory, but we perceive it as if it arrived according to a rectilinear path tangent to the real ray in the point where it reaches us. We would not know how to do it, but Eddington, in 1919, had the idea, knowing that on May 29 there would be a solar eclipse. Let us briefly describe what happened in the next reading.

---

[12] We refer to the so-called superior mirage.

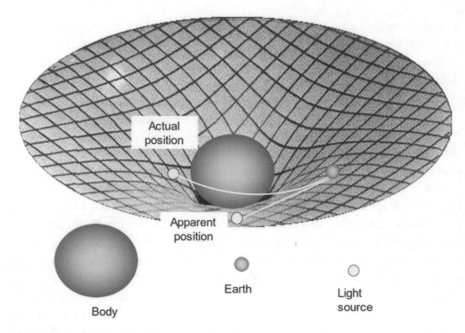

Fig. 5.28 Deformation of the path of a light ray due to the presence of a body. A light source (real position) is hidden from the earth by the body. Because of it the ray curves and, from the earth, it is seen as if the ray came from what is indicated as the apparent position

The first experimental confirmation of Einstein's conjecture

As we have said, this confirmation was first obtained by the astronomer Arthur Eddington, director of the Cambridge Observatory, shortly after the end of the First World War. Eddington had been informed of the new theory by a Dutch astronomer, de Sitter, who had sent him a memoir by Einstein on general relativity in 1916, in middle of the World War. Eddington's idea was simple and ingenious. The closest large celestial body to the earth is the sun. Of course, in its presence it is not possible to see any star, but we see it well at night when the sun is not on the earth–star trajectory. In the case of a total eclipse of the sun, it is on this trajectory, but its luminous disc is "shielded". If we refer to a star, we will see it in a certain position at night, but in a different one during the eclipse because of the modification of space due to the presence of the sun. Eddington organized two simultaneous expeditions, one to Brazil and one to Principe Island off New Guinea, where he went personally. He impressed photographic plates during the eclipse and at night and verified the existence of the expected deflection of light. Eddington finally estimated this deflection at 1.7 arc seconds against the 1.85 predicted by Einstein's theory. Considering also the characteristics of the equipment used, this was a sensational result that gave prestige and fame to the German scientist. Various experiments carried out over time have demonstrated the validity of these conclusions.

But how did Einstein react to the success of the experiment? Isaacson recounts that he was updated about the success of the expedition by Lorentz. He was talking with a doctoral student, Ilse Schneider, to whom he handed the cable with the news. The woman was enthusiastic, while Einstein, calm, said: "I knew the theory was right", to which she asked what he would do if the theory proved to be wrong and was told: "In that case I would feel sorry for the good Lord; the theory is right". To Planck's compliments, however, he replied more humbly: "It is a gift of benign fate that I have been allowed to have this experience". There was also a frivolous exchange of verses (whose poetic qualities are doubtful) between the scientist and his closest friends in Zurich. Solemn was instead the presentation of the results at the joint meeting of the members of the Royal Society and the Royal Astronomical Society in November 1919, gathered in a room of Burlington House in Piccadilly in the presence of a large portrait of Newton. Thus, Whitehead recounts: "*The atmosphere of tense and throbbing interest was that of Greek drama. We were the chorus commenting on the decrees of fate...and, in the background, the portrait of Newton, reminding us that the greatest of scientific generalisations was now, after more than two centuries, receiving its first modification*".

On one more thing, we want to dwell. A doubt that we could almost call classical is often posed in the following way: "light consists of particles called photons that have zero mass, how does any gravitational action exert itself on them?".

If you have followed what we have said so far, such a doubt does not really have any reason to be. The photon is also called quantum of light a kind of discrete packet of electromagnetic energy that sometimes seems to behave like a particle and others like a wave. Like others, it is a particle without mass so it is not subject to Newtonian-type forces. By now, we have understood that masses modify space and trajectories follow its shapes. Even the photon that has its own energy moves in this way. It is as if we were playing billiards on an undulating plane. The marble, to which we have given a certain energy, follows its depressions.

References

1. Isaacson, W. (2017). *Einstein—His life and universe* (pp. 118–126). Arnold Mondadori Editore S.p.A (in Italian). Translated from: (2008). *Einstein—His life and universe*. Simon.
2. Einstein A. (1905). Zur Elektrodynamik bewegter Körper. *Annalen der Physik, 17,* 891–921. English translation [PDF]

-    INFN   Rome   www.roma1.infn.it/.../A.%20Einstein%20-Sull%27elettrodinamica%20dei%20corpi%.   Accessed   September 2021.
3. Russel B., The ABC of relativity, TEA DUE, Tascabili degli Editori Associati, Milan, first edition 1997, p. 41 (in Italian). Originally published in English in 1925 republished as ABC of relativity by Routledge, Abingdon, U.K., 2009.

6

The Time of Man and the Time of the Earth

6.1 Gaia the Living Planet

So far, we have made a brief tour of various ideas of time. And we have seen that it is even possible to conclude that time with its objectivity and absoluteness does not exist. Einstein himself, in a letter he sent to the family of his friend Besso, after his death, said: "*For us who believe in physics, the division between past, present and future has only the value of a stubborn illusion*". Let us not talk about the subjective perception of time, on which we did not dwell. Let us now return to earth, in a literal sense. Let us deal with a different perspective, but for us humans fundamental, of how much space and time influence the nature in which we live and of which we are an integral part, but often not aware.

In this regard, we know that the various living beings have their own characteristic time, measurable at least by the average life span. It is a time that affects them not only individually, but also as a species. Let us go back to the macroscopic concept of time, the same concept of thermodynamics that so much conditions the destiny of systems, with an irreversible direction in its evolution (arrow of time) and bringing, moment by moment, new information and, therefore, often unpredictable changes. We also know that any living system is "open"; that is, it interacts with other systems in a more or less complicated and "entangled" way and each of its elements interacts with the others at different levels and with different intensity and speed. The systems existing on earth at the same time, in the same terrestrial time, as well as the single elements that compose them, are entities in themselves. Each one is

W. Grassi, *The Challenges of Time*,
https://doi.org/10.1007/978-3-030-94372-1_6

affected by the limits of its own sensorial capacities and by specific times and modalities of response to stimuli and interactions. Their existence takes place on different spatial and temporal scales on which depend also the modalities of relation both among different populations and species[1] (interspecific relation), and among individuals of the same population or species (intraspecific relation). There are various types of possible relationships among species. Here we will recall, because they will be useful in the following: competition, predation and mutualism. It is interesting to note what someone says about the latter in the case of two populations. With mutualism both benefit from what we humans might call collaboration, both in relation to their survival and growth. It is a bit like "union makes strength" and the opposite of the saying "mors tua vita mea" (your death is my life), which is more reminiscent of direct predation, one animal eating the other, and competition. Yet both exist and seem necessary, perhaps it is just a matter of a sense of proportion.

In 1979, Oxford University Press published a book by James Lovelock "Gaia: New Ideas on Ecology", which revolutionized ecology by proposing, as a scientific hypothesis,[2] the vision of the earth as a whole, as a single living system. It is, as we read in the description of the text later published by Boringhieri: "*a single living organism capable of self-regulation and of responding to all those new and adverse factors that disturb its natural equilibrium. Living matter does not remain passive in the face of what threatens its existence: the oceans, the atmosphere, the earth's crust and all the other geophysical components of the planet are maintained in conditions suitable for the presence of life precisely thanks to the behavior and action of living organisms, both plant and animal. Lovelock thus offers an alternative to the conceptions of those who see nature as a primitive force to be subjugated or conquered; or those who consider the Earth a spaceship gone mad, spinning aimlessly in the cosmos*".

It seems to be a system that, even though it sees a certain amount of competition acting within it, it mainly benefits from a sort of mutualism among the various subsystems. Since then, Lovelock's theory, completed in many subsequent publications, has become a fundamental element for a better understanding of the Earth's environment as a whole. By adopting this vision, it is possible to attribute to the Earth a unique time, a sort of its own time exactly as we attribute it to us, even though we are made up of distinct elements (from organs, to cells, to bacteria) with their own time scales.

---

[1] The species can be defined as an ensemble of interfertile individuals or potentially such.

[2] The idea that the earth was alive is an old one, as we have seen in the second chapter, but here it is proposed as a scientific hypothesis. It has, therefore, a quite different relevance.

James Lovelock's Gaia Theory

Like all the theories, also this one has availed itself of the contribution given in different times by various scientists, who, not always, knew one the studies of the other. Lovelock himself, born in 1919, names several of them, of whose existence he admits not to have known, at least at the beginning of the elaboration of his theories, as he says in the introduction of his book "The new ages of Gaia" [1]. His very short biography on the back cover of the book defines him as a doctor, biophysicist, chemist and inventor who worked for the National Institute for Medical Research in London and for NASA. It was at NASA that the hypothesis of the earth as a living organism took shape. At that time (1976), the Viking robots landed on Mars and communicated the news of the absence of life on the planet. The solar system was eventually completely barren. Only the earth stood out. Lovelock highlights, in the introduction of the book mentioned above, how they began to look at the planet with new eyes and found new things, among others: "*the radiation by the Earth of an infrared frequency characteristic of the anomalous chemical composition of its atmosphere*". He adds: "*I shall endeavor to show in the following chapters that unless life takes possession of the planet on which it has arisen and occupies it extensively, the conditions for its survival are not fulfilled*".

What does this prove? The Earth system developed in the past through interactions with matter (meteorites) and energy (solar radiation) from outside its boundary. But not only this. Its components (rocks, oceans, bacteria and plants) have also interacted and interact, organizing from within complex structures that have evolved and are evolving, always relating to each other and to the outside of the system. The time elapsed, past, to reach the present condition, present, has been infinitely long and the future depends on the possibility of maintaining these interactions.

In order to explain what Gaia is, Lovelock uses concepts typical of the science of complexity and thermodynamics, which we have already discussed in this volume and whose importance we have tried to demonstrate so far. In order to understand what Gaia is, that is, the entire earth conceived as a living being, it is necessary to first dwell on the concept of life. He says: "*Life is a social phenomenon. It exists in the form of communities and collectives. Physicists have a very useful term to describe the properties of groups: "colligative". It is a necessary term because there is no way to measure the temperature or pressure of a single molecule. Temperature and pressure, physicists say, are colligative properties of a sufficiently large group of molecules. All groups of organisms exhibit properties that are impossible to deduce from knowledge of a single organism in the group* [rejection of reductionism to acquire knowledge of systems]. ... *But what about large entities like ecosystems and Gaia? It was necessary to observe the earth from space ... to give us the sense of a real living planet, on which the living organisms, the atmosphere, the ocean, the rocks, come together to form the unique whole that is Gaia... Just as the shell is part of the snail, so the rocks, the atmosphere and the ocean are a part of Gaia. Gaia, as we shall see, existed in the past from the origin of life and extends into the future as long as life exists, Gaia, as a total planetary organism, has properties that are not necessarily recognizable through the observation of individual species or populations of organisms living together. In the Gaia hypothesis, as formulated by us in the 1970s, the atmosphere, oceans, climate and crust of the Earth are maintained suitable for life by the behavior of living organisms. Solar energy enables conditions suitable for life to be maintained. These conditions are stationary only in the short term and evolve in synchrony with the biota* [set of organisms (plants, animals, etc.) present in the ecosystem]. *Life and its environment are so closely linked that evolution concerns the whole of Gaia, not organisms and environment taken separately"*.

As we discussed in Chap. 4, we are talking about a living system, open to the contribution of solar energy, which constitutes the essential exchange for its maintenance, formed by an infinity of subsystems that behave according to their own rules, but not autonomously, having to adapt to interactions with others. At least with the closest ones. And here we must dwell on the double meaning of the adjective proximate. It is said in fact in the sense of space with the meaning of at a short distance, but also meaning that it is about to happen or has happened recently, that is, close in time. The concept is always the same, but the meaning of near and far has changed for man. In the first half of the 1800s, it took fifteen hours to make the journey from Bologna to Florence; with stops at post stations, today it takes half an hour. For a Bolognese, Florence is close because it takes a short time to reach it. A wayfarer or an animal, even if the roads have improved, still takes a time similar to that of the nineteenth century. For a man, travel times have changed thanks to technology, but certainly not for animal species, unless they are able to exploit it as certain insects do.

They are systems with their own spaces and times, as well as the relationships that bind them. They are times according to which they are formed, evolve and react and are often very different from each other. In some cases, they are such that some do not even realize that the others are also changing. It is enough to think of the earth, with its continuous morphological modifications that we are essentially aware of when it manifests them in a violent way with earthquakes and eruptions. Or the oceans with their changes in temperature, salinity and currents, which have had to highlight phenomena such as the melting of the ice for the common man to become aware of them. On the opposite side, there are, for instance, the ephemerals whose adult life lasts about one hour and a half, at least according to our time of great terrestrials. What do you want an ephemeral to know about the world if it has spent its life as a larva in a pond and as soon as it leaves it does so to reproduce and has so little time?

Environmental conditions and resources are part of Gaia. Certain organisms can modify some of them, sometimes only on a local scale and sometimes, as in the case of man, even on a global scale because of the number and invasiveness of his activities. And this is a crucial point: the human species has a power that is immeasurably greater than that of any other. I would say that, for Gaia, it has long since begun to constitute a pathology. Linked to the environmental conditions are the resources, which are consumed to directly favour the growth of the individual or to have the energy necessary for his development. Even a child understands that they are limited, but man has worked a miracle and established that his well-being increases with the growth of their consumption; only someone, with a little more modesty, will speak of their use.

It must be said that Lovelock's position has changed over the years and a 2016 article by Francesca de Benedetti in the newspaper La Repubblica [2] summarizes the changes well. In 2006, with the book Gaia's Revolt, he claimed that within the current century the earth would be inhabited by only a handful of humans due to anthropogenic production of CO_2.

In the eye of the article [2] cited instead reads: "*Climate change? "Not unsustainable". New technologies? "They will change us more than the planet". Robots? "As long as they don't vote". The animal of the future? "The electronic one, a fusion of man and chip. To him one day it will seem like two thousand years". The word to the 97-year-old scientist who developed the Gaia theory and predicted the end of humanity. But now he has changed his mind"*.

In particular, we refer to the latest book, a collection of essays by various authors, edited by Lovelock "The Earth and I" published by Taschen in 2016. Again according to the article, it is an injection of hope, of confidence that man "the most extraordinary animal" can save himself. Pathology, according to this approach, gives signs that it can fall within the realm of physiology.

I have not read the book, but I read the newspapers, follow the news and find it hard to share that hope. Among other things, Lovelock cites the 2015 Paris Agreement saying that it showed that we are capable of taking the climate issue really seriously, though he warns that everyone will have to revolutionize their behaviours. So far, it does not look like things are going in that direction and no revolution of human behaviour seems to take place. In 2018, Fatih Birol, executive director of the International Energy Agency (IEA), said he expected global emissions to rise for the second year in a row. In addition, he pointed out that there is "*a huge amount of coal-fired, very young power plants*" in Asia that are likely to remain in operation for another forty years or so. In November 2021, a conference (COP26) was held in Glasgow for urgently committing countries to reach to cut their emissions to net zero by 2050 (planned by Europe and USA). India and China, respectively, pledged to reach this target by 2070 and 2060, in agreement with the above projections. Once again, man is promising and Gaia is supposed to take his promise despite not knowing for how long.

Trump's enlightened vision, and not just his, everyone knows. And we could go on like this for a long time. The new White House tenant (J. Biden) said, at COP26, that every day the world delayed in tackling climate change, the cost of inaction increased. And Gaia should tolerate this up and down. We could go on like this for a long time.

Does time also enter into this? Certainly, but it is a time seen from another perspective. Meanwhile, let us try to define a common time referring to Gaia, independently of the various calendars adopted by the different human populations. The origins of the earth are traced back to four and a half billion years ago, and we can arbitrarily take this (but we could act in many other ways) as the "instant of birth", just as we calculate our age from the moment we saw light. Inorganic and living forms then developed progressively in complex ways that we will not discuss in detail here, but of which we have tried to give an idea. Today, if we restrict our view to the human species, which is the only animal species capable of heavily conditioning the life of Gaia, we see a series of peoples who are each at a different evolutionary, socio-economic and cultural stage. It is as if each of them were living their own historical moment, therefore their own historical time, often very different from that of others. The history of each was first born in an autonomous way and progressively modified by interactions with different peoples. Gradually, these interactions have extended to greater and greater distances and with shorter and shorter times. As if space–time on earth, for the human species, had shortened. In the meantime, the rest of Gaia was only a stage on which man acted, taking it into consideration only to overcome its resistance or suffering the effect of those natural catastrophes, long commonly interpreted as punishments to man by an angry god.

Unfortunately, man has often focused on himself as a separate entity from what surrounds him, uncritically following his way of thinking, considering himself more the master of the house than an integrated being who participates in nature. He behaves just like the manager of a house that is not his own, living an independent daily life and, almost always, doing the necessary "maintenance" only when he cannot do without it. In the meantime, I do what I like and, even if I imagine that there are damages, I do not want to see them, or, if I am not directly interested, I do not care about them at all. The examples of indifference are endless, from whale hunting,[3]

---

[3] Which still continues in Japan, despite the Hague court's ruling.

to strip-mining in the Appalachian Mountains,[4] to the killing of rhinos for their aphrodisiac horns or elephants for the ivory in their tusks. And as time goes by, the damage accumulates and pieces of Gaia die.

The logic of robbery of natural resources, with great consequences also for the human species, is an old one, and indifference to it too.

By now, two time scales are clearly emerging: that of Gaia, who needed a long time to get where she is, and that of man, with incomparably short times. It is a bit like the last heir of a family, who has built up his heritage over generations, busy squandering it all on a poker night.

6.2   A Simple Interaction Model

We could say that, while the human species is able to control the quantitative and qualitative depletion of resources, it is Gaia, with its infinitely longer timescales, that controls their renewal. This is true for all environmental resources, whether they are fossil or not.

How do space and time play into this view? Space is confined by the boundaries of the earth, the thickness of the atmosphere is obviously included. Matter, living or not, is confined in this space, by necessity. Everything is maintained by the interactions that take place within it. The only effective interaction with the outside of this system is the radiant energy that comes from the sun and which the earth sends back into space. The most immediately visible effect today is precisely the alteration of the balance of this interaction. It is also called the greenhouse effect. Why? A greenhouse is a structure made of transparent material that lets in the solar radiation emitted at high temperature and therefore at a low wavelength, particularly in the visible range. On the other hand, it retains part of the radiation sent towards space by internal bodies at room temperature and therefore at a high wavelength (infrared). If we want to change these two contributions, all we have to do is modify the properties of the transparent material. And, this is what we do with the gases and dust we send into the atmosphere. This is only one factor of perturbation, but it has serious effects on the balance between the elements of Gaia.

Here again, time is crucial. It took an eternity for the earth, with its temperatures, climates, animals and vegetation, to reach the condition in

---

[4] Coal companies flatten mountains with explosives to unearth ore seams. The peaks are cut by the explosions, the debris, thrown into the valleys, creates dams full of toxic substances that divert the watercourses and pollute the water tables, beautiful forests are completely destroyed. Among other things, this is closely linked to the emergence of a high number of cancers and other diseases in the area.

which we older people found it. It has thus reached very precise values of the physical quantities and percentages of the substances that characterize it today. Its equilibrium is based on very precise values of certain quantities. If they are changed too much, they lead to very serious consequences. We will mention only two of them, whose significance is easy to realize, but there are countless others. The human body must remain at a central temperature, the one measured with a thermometer, of 37 °C with an oscillation of half a degree. Half a degree is enough to pass from a physiological condition to a pathological one.

For substances, let us limit ourselves to two gases: oxygen (O_2) and carbon dioxide (CO_2). The first is fundamental for life and is at the basis of the oxidation process. Its volume percentage in the atmosphere is around 21%. But below 17%, man cannot breathe. Too high a percentage would put our (and not only) existence at risk. As Lovelock says, "*you can't light a fire, not even of dry branches, if the oxygen content is less than 15%; above 25%, on the other hand, the fires are so intense that even the damp wood of a tropical forest burns with a terrible conflagration*".

What about the much reviled carbon dioxide? Together with various other elements and phenomena, not all well understood to date, it has contributed to the current state of life on earth. To put it briefly, we can still make our own the words of Lovelock: "*Since the beginning of life on Earth, carbon dioxide has played a contradictory role. It is the food of photo-synthesizers, and therefore of all life; it is the intermediary through which the energy of sunlight is transformed into organic matter. At the same time it has always served as a blanket that has kept the Earth warm: a blanket which, now that the sun is hot, is beginning to become a little heavy, but which we must continue to wear, because it is also our means of sustenance*".

With infinite patience and times not commensurable with human existence, Gaia has brought itself to its present conditions. Its evolution has been very slow, but it is raising "a snake in its bosom": a species that has reached a number of individuals and technological capacities such as to be able to alter both these conditions and their evolution in a very short time. It is a species that is often unable to foresee all the consequences of its actions or to cope with them in the long term and for it the saying "close the stable when the oxen have bolted" applies. We should also keep in mind that most phenomena do not behave linearly (concept we are so much attached to), but they have thresholds beyond which they change abruptly. In addition, this change is not always predictable and affects all the others it interacts with.

It is the time to take a look at what scientists have said about the relationships between species, a subject that we will deal with in particular as relationships between the species man and his environment. To do this, we will limit ourselves to a few essential elements. The model we are referring to is the one that goes by the name of Lotka (1880–1949) and Volterra (1860–1940) and that dates back to the period between 1925 and 1926. We will use it just to provide some ideas. It refers to the interaction between two populations: one of prey and the other of predators. The growth of both depends on the number and fertility of the individuals and on the resources available to the prey, assuming that the only resource for the predator is the prey under examination. At first, we also assume that prey have unlimited resources available to them. We begin by seeing how their numbers vary over time. In the absence of the predator, the number of prey would grow infinitely, more or less rapidly depending on their fertility. If n is their number, A_n ($A_n = N_n - M_n$) the growth rate,[5] equal to the birth rate, $N_n(>0)$, decreased by the mortality rate, $M_n(>0)$, the variation Δn of the individuals in a certain time interval Δt is related to the number of individuals present at time t. The growth rate has a positive value if the birth rate exceeds the death rate ($N_n > M_n$); otherwise, it is negative. If $A_n = 0$, the number of individuals remains stationary. There is no change in the number of components of the population. In mathematical terms, using a continuous formulation, we write:

$$\frac{\Delta n}{\Delta t} = A_n n$$
$$A_n > 0 \text{ if } N_n > M_n$$
$$A_n < 0 \text{ if } N_n < M_n$$

To learn more, we must also specify what is the number of individuals, n_0, present at the beginning of the observation, that is at a time that we can comfortably assume equal to zero: $t_0 = 0$.

Let us make an example on how to use the above equation. For 2009, estimates of the growth rate of the world's human population were 1.1% per year, with a birth rate of 1.9% and a death rate of 0.8% over the same period. This resulted, in the year, in an increase of about 75 million people. Referring back to the formula above the term Δn is worth 75 million, $\Delta t = 1$ year and $A_n = 1.1/100$. To calculate the number of individuals, n_0, present at the beginning of 2009, we can put the quantity n to the right of the equal $n = n_0$. It can be obtained from the previous data with some simple steps:

$$\frac{\Delta n}{\Delta t} = \frac{75 \times 10^6}{1\,(\text{year})} = \frac{1.1}{100} n_0$$

---

[5] If a population at the beginning of the year is of 100 individuals and at the end of 110 individuals, the growth rate is 10% per year, with respect to the population present at the beginning of the year. We have to keep in mind that the growth rate of a single population is linked also to the incoming and outgoing migratory phenomena.

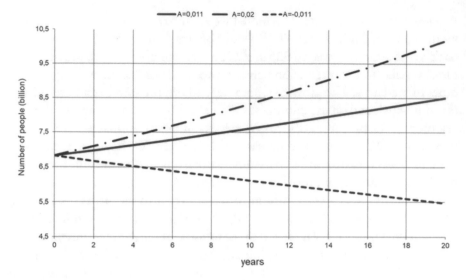

Fig. 6.1 Population change as a function of growth rate

$$n_0 \cong 6.82 \text{ billions}$$

Solving the equation (differential if we go to very small, infinitesimal incre-ments[6]), we find that at a constant growth rate the number n of individuals varies exponentially as in Fig. 6.1. In the graph, we start with data from 2009 and examine two positive growth rate values ($A_n = 0.011$ and 0.020) and one negative growth rate value ($A_n = -0.011$).

Despite not having unlimited resources, in the presence of various causes affecting the rate of growth other than predators, the number of individ-uals of the human race has nevertheless increased. In fact, some estimates of the world population give the following values, in millions, from the year 1000 A.D. to the present time.

Year	Population (million)
1000	400
1500	458
1600	580
1700	682
1800	978
1900	1.650

(continued)

---

[6] We still refer to continuum hypothesis thanks to the large number of individuals.

(continued)

Year	Population (million)
2000	6.070
2010	6.930
2015	7.349

These figures are eloquent in themselves. They certainly show an exponential growth from 1900 to the present day with an increase of more than five and a half billion in little more than a century with an increase of 450% compared to 1900. In this regard, Enzo Tiezzi[7] (1938–2010) mentions an anecdote [3] that we report in its entirety: "*I remember a seminar in two voices, with Father Ernesto Balducci, at the Cittadella in Assisi. A young man objected to me, about the problem of the demographic increase, that in the Sacred Texts it was written "grow and multiply". Ernesto touched my arm and made me sign that he would answer in my place. "Read well!" he said to the young man, "it is written grow, multiply and fill the earth. Today the planet is full!""*.

We made the assumption that resources are unlimited. Otherwise, what happens?

Let us examine the case of a natural park, which we assume is populated only by deer, where there are no predators. This case is the same as the previous one, but with finite resources. In a limited space, the vegetal resources are also limited. The fauna, in this case, acts as a predator for the flora. Both reproduce according to different speeds. The number of deer present must be such that the vegetation they use as food has time to regenerate. There is, therefore, a maximum number of individuals that the ecosystem can support. This is called the "maximum carrying capacity of the system". More significant is the so-called optimal carrying capacity, which is the capacity that the ecosystem is able to maintain in the long term in the presence of random environmental variations that may occur. Of course, it is related not only to the number but also to the type of individuals and their behaviour. The times of consumption of the vegetation must be commensurate with those of its regeneration and, together with the available space, determine the maximum value of the number of animals sustainable in the long term.[8] It does not take much to understand this, but doing so is a whole other matter.

---

[7] Professor at the University of Siena and famous environmentalist, he has collaborated with Barry Commoner.

[8] We haven't mentioned it, but water is obviously a fundamental resource for any form of life.

Otherwise, the growth rate cannot be kept constant, both because the mortality rate of the young may increase and because the birth rate is reduced.[9] In the park, the number of individuals is kept under control, either by transferring them elsewhere or by programmed selective culling. This creates the same effect as the predator. If this does not happen, the animals leave the limits of the park and go to get food in cultivated and/or inhabited areas. Invasions of wild boar or even of the more timid fallow deer in the countryside are the order of the day. On the other hand, everyone knows about the migration of elephants or wildebeests in Africa. The animal is always looking for the most favourable environmental conditions or expanding the area of influence (park + countryside) or leaving a zone to move to a more favourable. If there are predators, they also try to move away from places where their concentration is too high. On the other hand, also the men have always done it and also my country (Italy) has witnessed it with the emigration towards USA, Argentina, Germany, Belgium, etc. And migration has also taken place within borders: from the agricultural South to the industrial North. Today, the phenomenon is becoming much more global, even if it is inadequately addressed. I remember a saying, I think from the United Nations, "don't bring them fish, but teach them to fish". It does not seem to me that this has been done with the necessary effectiveness.

But these are only redistributions of individuals within the Earth system and man has no chance of getting out of the park, as deer do. The available space is limited and the time to provide for not saturating it as well.

Let us now dwell on the effect of the predator, again assuming that the prey still has unlimited resources. The only resource the predator has is the prey. The quantity of food available depends on the number of "useful" encounters with them. "Useful" because only a percentage of the encounters between prey and predator are successful, in the sense of providing food for the latter. The parameter that gives the fraction of these encounters is indicated with S, called rate of predation. The predator also needs a minimum amount of food to survive. For each useful encounter, only a fraction C (prey–predator conversion rate) contributes to this.

If, as said, n indicates the number of prey and p the number of predators present in a certain delimited territory, the number of possible encounters is given by their product np of which Snp are useful. The contribution to the formation of another predator is, finally, $CSnp$.

If the predator finds no prey (absent prey, $n = 0$), its growth rate, A_p, can only be negative so its number declines and it dies out, unless it moves

---

[9] Wild life tries to adapt to the contingent situation. The same births occur at times when resources are most abundant. The number of newborns is also regulated on the basis of these.

elsewhere. Obviously, while the preys decrease for each useful encounter by *Snp*, the predators increase by *CSnp* (the number of preys must obviously be greater than the number of predators). With the same previous meanings and, indicating with the subscript n the quantities related to the prey and with p those referred to the predators, we obtain the following relations, starting from the initial values (at time zero, $t = 0$): n_0 and p_0.

$$\frac{\Delta n}{\Delta t} = A_n n - S \cdot n \cdot p$$

$$\frac{\Delta p}{\Delta t} = A_p p + C \cdot S \cdot n \cdot p$$

$$A_n > 0; \quad S > 0; \quad C > 0; \quad A_p < 0.$$

From these, we return to the previous ones in the absence of predators ($p = 0$) and the reduction of predators in the absence of prey ($n = 0$). We easily derive the value of the equilibrium condition, in which both the number of prey and predators are kept constant (the variations n and p with respect to time t are nil).

$$0 = A_n n - S \cdot n \cdot p$$

$$0 = A_p p + C \cdot S \cdot n \cdot p$$

$$0 = A_p p + C A_n n$$

$$\left(\frac{p}{n}\right)_{\text{Equilibrium}} = -\frac{C A_n}{A_p}$$

Keep in mind that A_p is negative. With this parameter being equal, the equilibrium value increases the higher the growth rate of prey and the prey–predator conversion rate. In the case of predatory man, at first he had to move around in search of prey and other products that nature made available to him. This was the phase of the hunter–gatherer. As also happens with animals, especially during the period of reproduction, between groups that had the opportunity to insist on the same territory, competition was often bloody.[10] In becoming "civilized", the human being discovered agriculture and breeding. He no longer needed to spend a good part of his time moving around in search of food. In time, he identified and selected species that he could cultivate, breed and use for his own purposes, including food. For the animals used as food and, therefore, prey, we approach the condition in which the resource for prey is unlimited thanks to the breeder. He himself

---

[10] It was the beginning of wars for hoarding resources.

is the direct predator (who physically kills the prey), but the real predator–prey relationship, indirect, moves to the supermarket between customer and "goods" on display. The average person no longer needs to pull the neck of a chicken or a stick to kill a rabbit. He no longer has the perception of acts like these just as he no longer needs to collect wood or fetch water from a spring. Fortunately, you will say and I will agree. But losing the perception of these acts, the perception of the time needed to carry them out and the fact that, far from us in the civilized world, there are still those who have to do these things on a daily basis (fat chance if they have the chance), distances us more and more from certain realities. We do not have the need to get the resources ourselves and, therefore, to touch what it entails, nor to understand the implications it has for those who still have to do it.

Getting back on topic let us stop with an example. If man is the predator and chickens are the prey, try looking on the Web at some pictures of how to achieve the maximum value of predator–prey ratio on intensive farms, which, said otherwise, means churning out as many chickens as possible in order to satisfy the maximum number of customers. The direct predator, the farmer, is also the one who promotes the birth rate and determines the growth rate of the chickens by minimising their life span and treating them no more and no less than objects in mass production. So, A_n reaches the greatest possible value, even if there will always be random events that will not allow it to be kept strictly constant. Of course, every predator–prey encounter is useful ($S = 1$) and C is maximized by bringing the animal to an optimal weight in an equally optimal time. From the point of view of productive logistics, everything has been thought of. It seems that in Italy are "bred" (quotation marks are a must because it is more of a production chain in most cases) about half a billion chickens per year of which only 5% in non-intensive farms. This leads to a density of animals that can even exceed twenty individuals per square metre. Reducing growing times and available space is imperative with an increase in human population such as the one we have been talking about. It is a sector in which "increased production" will always be required, other than the ups and downs of the automobile market. Obviously, natural life spans have nothing to do with it, ranging from 35 days in intensive farming to 120 days where the chicken is given access to free outdoor space. Biological times are eliminated, and time is given by man, as, unfortunately, in many other cases. We could say that man replaces fate. The difference is that for a breeding farm man is able to condition the birth rate of individuals, while for many other situations, he seeks to maximize profit, as always, but can only control the consumption and, therefore, the mortality of the resource.

Even if he recycles, a share of waste is ineradicable as are the traces on the environment.

An Example of Human Impact

To complete what we have said and to emphasize how even aspects, which seem harmless because they are habitual, in fact are not, I report below a summary of what Lorenzo Brenna says in an article published in "Lifegate" on May 25, 2018, [4] entitled "Humans constitute only 0.01% of life on Earth, but have exterminated 83% of wild mammals". The article refers to a study led by Prof. Ron Milo of Israel's Weizmann Institute of Science who writes, "*I hope this study will provide people with a perspective on the dominant role humanity now plays on Earth*".

The study carries out an analysis of the distribution in terms of biomass of the planet's living organisms, including viruses. It was concluded that plants make up 82%, bacteria 13% and all other creatures (insects, fungi, fish, animals, etc.) only 5%. From the same point of view, the human species, consisting of more than seven and a half billion individuals, is 0.01%. Nevertheless, our impact on the rest of the living world has been and is devastating. It has progressively depleted the Earth of wild plants and animals, replacing them with livestock. Since its appearance, "homo sapiens" has caused the extinction of 83% of wild mammal species and half of all plants.

The drastic decline in wild species is matched by a sharp rise in farmed livestock. It is estimated that poultry account for 70% of all birds, while mammals are even worse off. 60% is accounted for by farmed animals, 36% by humans and only a measly 4% by wild animals.

Biological diversity is being depleted with increasing speed. Biodiversity, the wealth of our planet, is decreasing frighteningly.

With a sort of logical leap, but not unmotivated, we can say that by reducing the variety we reduce the number and type of interacting subsystems within the great Earth system. We thus profoundly alter a series of relationships of which we do not yet know all the aspects and, therefore, not even the consequences of their suppression. A thermodynamicist would say that we are causing a dangerous increase in entropy.

Milo says: "*It's definitely amazing how disproportionate our impact on the planet is. When I do a jigsaw puzzle with my daughters, it usually features an elephant next to a giraffe next to a rhino. But if it were more realistic it should depict a cow next to a cow next to a cow next to a chicken*". He adds: "*Our **food choices** have a big effect on the habitats of animals, plants and other organisms. I hope this study will help people to revise their view of the world and how they consume*".

With regard to livestock farming alone, it is essential to underline the enormous impact it has on the production of greenhouse gases, according to some more than the entire transport sector, due above all to the methane released. Not to mention water pollution, deforestation and other aspects directly or indirectly connected to it.

This is a fundamental and decisive point. Above all because the decisions taken and the actions performed today will have repercussions on the destiny of a more or less near future of Gaia and of our own species. The words of J. F. Clarke (1810–1888), American preacher and theologian, come to mind: *"A politician looks to the next election; a statesman looks to the next generation. A politician thinks of the success of his party; a statesman of that of his country"*. We would already be happy to have politicians who think of their country, and therefore of a much wider area than that confined to their own person. However, the vision must be much broader, both in space and time, if we really want to try to deal with the problems on the table in an organic and concrete way.

Knowledge and open and transparent collaboration are the basis from which the decision-makers of the fate of the world must start. If, thinking about today's situation, it seems a joke to you, I can only agree, unfortunately!

Returning to Lotka and Volterra's model, another term is added in it, when the finiteness of resources must be taken into account. This condition, in fact, determines a competition among the preys (fallow deer have a limited vegetable resource available) and among the predators because the number of preys is also limited.

$$\frac{\Delta n}{\Delta t} = A_n n - S \cdot n \cdot p - \left[F n^2 \right]$$
$$\frac{\Delta p}{\Delta t} = A_p p + C \cdot S \cdot n \cdot p - \left[B p^2 \right]$$
$$F > 0; \quad B > 0.$$

As usual, we do not solve the equations, but we observe that the terms added and put in square brackets depend on the square of the number of individuals present. The number of available individuals is, therefore, decreased over time by intraspecific competition and is much more sensitive to the effect of competition than to other factors. In the absence of predators ($p = 0$), the increase in time of prey is:

$$\frac{\Delta n}{\Delta t} = A_n n - \left[F n^2 \right] = n(A_n - F n)$$

The individuals of a species can stop increasing ($\Delta n / \Delta t = 0$) if $n = A_n / F$. If the intraspecific competition is strong (nF greater than the growth rate), the number of individuals decreases. With all the cautions of the case, since every model is rigorously valid only in certain areas, we can say that the numerosity of the individuals is linked to the competition for the resources. The more difficult it is to find them, especially if they are distributed with a density that is difficult to increase in a limited space, the greater the level of competitiveness. Starting from Cain and Abel, the human kind does not need to demonstrate its potentiality of competition. History also teaches us this by adding the real motivation of hoarding riches of all kinds and increasing power, without any regard for the lives of different populations of the same species. Now they were inferior races; now they were godless ignorant savages and hypocritical justifications for certain actions flourished and flourish, assuming one felt or feels the need for them. And in the past, there were few of us and the resources were many, albeit reachable with difficulty. Now we are many and the resources are few, I don't dare to imagine to what point the competition among the peoples will push itself. And the conscience of the governors? There is a very recent example, perhaps not the worst, but it would certainly be farcical if it were not tragic. In August 2019, the President of the USA, having learned of the melting of the ice, had an "enlightenment on the way to Greenland". What could be better than ice-freeing a vast region rich in natural resources? Not being a prevaricator, he offered to buy it. However, he found the "Denmark shop" closed and, disappointed by the service, he postponed his official visit to Denmark after having been invited by the royal family for September 2nd and 3rd. But that is not enough, at the same time a part of the Amazon, the lung of the world that produces 20% of the oxygen of our planet, is burning due to arson. Only after threats of international sanctions did Brazilian President Bolsonaro, always accused of promoting a policy of deforestation, decide to use the army. In the meantime, a global disaster is being perpetuated because of a senseless local policy. I wonder what Lovelock is thinking?

And in the meantime, the world, deeply and appropriately sick of social media, willingly leaves to Greta Thunberg the task of saving it with the useful, as easily exploitable, naivety and enthusiasm of sixteen years old. After being received by the Pope, Obama and other greats of the Earth, she has been heard by the United Nations assembly (September 2019), who has been enlightened for a day on the problems that her concrete indifference has helped to create and increase. Many of us have tried in vain to walk similar paths day in and day out for decades, so I sincerely wish Greta all the luck and success she can have. Let us hope that her path will be more fruitful than ours, despite of her disappointment after COP26 in November 2021.

6.3   The Horror of Unconscious Human Stupidity or a Gaia Signal of Intolerance?

Whatever it was, human stupidity played a fundamental role. While I was waiting to get the drafts of the Italian version of the book before publication, originally scheduled for early May 2020, here comes a tiny (no bigger than 400 millionths of a millimetre) acellular microorganism, known as a virus and in particular by the name of COVID-19, spreading terror over the entire earth, directly affecting "even" the so-called industrialized world. It started in China and was gradually but surely spreading everywhere. We were (and maybe still are) in the midst of a nightmare that we have only seen at the cinema in so-called catastrophist films. We could not leave the house, if not for basic needs, only the essential activities were still in operation, people who must, went out with masks, outside pharmacies and the few shops still open there were queues with people at least 1 m away from each other. If you were walking on a sidewalk and someone else came up to you, one of you switched sidewalks. The police did checks to make sure you had a valid reason for being out of the house. They also used drones. The maximum distance you could be away from your home, unless you had a proven reason to be away, was 200 m. You could go out, with a special written and signed self-declaration, for serious health reasons, to go to the pharmacy, to go to work, only for some strategic categories, to do the shopping (with limited entry and in small grocery stores one at a time), to buy the newspaper, to take the dog out. Doctors and health personnel were required, which in some places were absolutely lacking. In some areas, the intensive care units were insufficient and field hospitals were set up. There was a well-founded fear that the lack of personnel and facilities could spread and that it could be necessary to decide who should be given priority in treatment and who should not. In some countries, politicians were talking openly about this.

Never has it been more true that one dies alone. In the last moments, if you were ill with COVID, you did not even have the comfort of the warmth of family members, nor could they have that of a final farewell. There were no funerals. In April, the newspaper La Repubblica reported that in New York (where on 10 April the number of cases of COVID's disease exceeded 170,000 and the number of deaths reached 7800) "*in such a situation, there is no other choice: to bury the nameless, the family-less, the penniless in the mass graves of Hart Island*".

It all began, it seems, in Wuhan a city of eleven million inhabitants capital of Hubei province in central China. From 18 to 27 October 2019, the military "Olympics" was held there, with the participation of around 140 countries and a total of over 10,000 athletes. Never a phrase of circum-stance pronounced by Hervé Piccirillo, President of the International Military Sports Council (CISM), according to which the games were a positive step to bring people together and that sport can send a message of peace and sharing, seemed more ironic in retrospect. The peace I do not know, but the sharing of the virus there was, although there was not a great unity of the states in dealing with the pandemic.

The first rumours said that the virus had spread from the fish market, but this is all to be verified. Regardless of any other interpretation, it is not difficult to see how the return of 10,000 people, some of whom were infected, to 140 countries contributed to the spread of the virus.

Already during the course of the games, five foreign athletes, whose nation-ality was not revealed, were taken to the City Jinyintan Hospital as they were subject, according to official statements, to "imported and transmissible infectious diseases".

As things stand, I believe there are three types of hypothesis that have emerged.

- According to two biologists from the South China University of Tech-nology, everything originated from infected animals (including 605 bats) kept in one of the two laboratories located at the Wuhan fish market and not from direct transmission from bats to humans, as officially claimed by Beijing. Among other things, a military exercise was held in Wuhan in September to deal with a bacteriological threat ahead of the military games.
- Chinese government officials, in response to comments from the USA, suggested that US troops brought the virus to Wuhan. They also pointed out that the US representative team finished only 35th in the competitions, which is strange for generally strong athletes.
- According to others, this is a real "spillover" or "interspecific jump". It corresponds to the passage of a pathogen from one species to another. When it infects a host organism, the virus mixes its genetic make-up with that of viruses already present or modifies parts of its RNA. It reproduces at the expense of infected cells and leaves the host with a different genetic make-up, which puts it in a position to propagate to other species. The case of the Wuhan market, where there are many animals of different species, is certainly favourable ground for such a process to occur.

It is not the subject of this book, nor do I have sufficient elements nor the specific competence to make precise comments or give a judgement. But time has something to do with it, and it is the one to which a univocal direction is attributed. It is the time for which a greater or lesser delay in dealing with problems can drastically alter their consequences. The various international conferences, with their substantial failures, have already shown us man's farsightedness and the concreteness and immediacy of the measures taken. Now COVID-19 has made itself felt directly on our skin and time is measuring it. In spite of this, mankind has reacted in a scattered order, putting in place its demented criteria concerning borders, rivalries between states and everything that has always guided its behaviour. The episodes are many. China's wicked delays in reporting the spread of the virus speak for themselves. Italy, one of the countries most affected right from the start, was initially mocked and isolated by various countries, including European ones, which then adopted the same restrictive health measures with great delay. So much so that the president of the European Commission Ursula von der Leyen said "It is right" that E.U. [European Union] apologizes to Italy for the lack of solidarity shown at the beginning of the pandemic, but "now Europe is there", speaking at the plenary session of the European Parliament on 15 April 2020. Initially, the USA had decided to stop flights with all European countries except UK. Perhaps Brexit [exit of UK from the European Union] was a sufficient health guarantee or were the hallucinating statements of Boris Johnson[11] [who later also suffered from the virus and got off lightly] on how to deal with the crisis, reported by an Italian newspaper [5]. And there are a lot of examples of the absence of awareness, solidarity and political sensitivity. In fact you are spoilt for choice, just take a trip around the Internet, taking a look at what happened in March 2020 and later on. Not to mention what some have called the "constant background noise" of certain so-called politicians who never refrain from speculating on anything and everything at any time to keep themselves "visible". Hopefully people will sooner or later understand the difference between a politician and a jackal, whichever side they belong to.

The usual blindness has persisted whatever the political colour. But mutual distrust has also played its part. Man knows that he is idiot enough to be at the origin of what is happening. We have arsenals of all kinds including those with bacteriological weapons and people who work in them and provide their knowledge and skills to refine the destructive technologies that are their ultimate goal. Maybe they do it to defend their country or their personal gain,

---

[11] " …Premier **Boris Johnson**'s statement on the ongoing **pandemic**, to be dealt with strictly without moving a finger. That is, the proclaimed "herd therapy", according to which the immunization of the survivors will be obtained after the inevitable extermination of the weakest. The demented declination of Darwinian survival as a pseudo-scientific justification for not wanting to **waste pounds** on the rescue of poor people. One step closer to humanitarian catastrophe after weeks of Coronavirus **denialism by** our colleague across the Atlantic: US President **Donald Trump**."

without any effect on the tranquility of their sleep. I knew this reasoning as a young man when I refused, if only to sleep peacefully, a job in a similar sector: "if you don't do it, someone else will". It is therefore logical that conspiracy theories have arisen, which I certainly do not feel like excluding.

But what if it is not so and it is the biosphere that every now and then gives us some sign of intolerance? It did it in the remote past (Plague, Smallpox, Cholera), in the recent past (Spanish Flu, HIV, Sars, etc.), and it is doing it now. Life on earth was born with bacteria, could it be that the life of the most invasive species on the planet will end with viruses? What is the time limit, what determines it? Do we have to wait for the changing of the climate to change the whole planet or is there something more surreptitious and insidious that in the meantime reacts to the excessive growth of a species?

David Quammen confirms this in his book [6], and we report directly his words "*Let it be clear from the start: there is a correlation between these diseases that pop up one after another,*[12] *and they are not mere accidents but unintended consequences of our actions. They are the mirror of two converging planetary crises: an ecological crisis and a health crisis. When added together, their consequences show up in the form of a sequence of new, strange and terrible diseases, emerging from unexpected hosts and creating serious concerns and fears for the future for the scientists who study them. How do these pathogens make the leap from animals to humans and why does this seem to be happening more frequently in recent times? To put it as simply as possible: because on the one hand the environmental devastation caused by our species pressure is creating new opportunities for contact with pathogens, and on the other hand our technology and social patterns are helping to spread them even more rapidly and widely. There are three elements to consider.*

One. Human activities are causing the disintegration (and I did not choose this word at random) of various ecosystems at a rate that has the characteristics of a cataclysm. We all know how this happens in broad strokes: deforestation, road and infrastructure construction, increased farmland and pasture, wildlife hunting (strange, when Africans do it it's "poaching", when Westerners do it it's a "sport"), mining, increased urban settlements and land consumption, pollution, the spilling of organic substances into the seas, the unsustainable exploitation of fish resources, climate change, international trade in goods whose production involves one or more of the problems described above, and all the other activities of "civilized"

---

[12] Here is the list of viruses that have manifested since mid-century 1900 that he cites: "*In a list of the topical and most anxiety-provoking moments of this saga, besides Machupo, one cannot miss Marburg (1967), Lassa (1969), Ebola (1976), HIV-1 (indirectly recognized in 1981, isolated in 1983), HIV-2 (1986), Sin Nombre (1993), Hendra (1994), avian flu (1997), Nipah (1998), West Nile fever (1999), SARS (2003) and the much-feared but ultimately not very serious swine flu (2009).*"

man that have consequences on the territory. We are, in short, crumbling all ecosystems. This is not a very recent development. Humans have practiced most of these activities for a long time, albeit for a long time with the help of simple tools. Today, however, there are seven billion of us and we have modern technology on our hands, which makes our global environmental impact unsustainable. Tropical forests are not the only environment under threat, but they are certainly the richest in life and the most complex. Millions of species live in these ecosystems, most of them unknown to modern science, unclassified or barely labeled and poorly understood.

Two. Among these millions of unknown species are viruses, bacteria, fungi, protists, and other organisms, many of them parasites. Specialists now use the term "virosphere" to identify a universe of living things that probably makes any other group pale in size. Many viruses, for example, inhabit the forests of central Africa, parasitizing specific bacteria, animals, fungi or protists, and this specificity limits their range and abundance. Ebola, Marburg, Lassa, monkey pox and the precursor to HIV are a tiny sample of what the menu offers, of the myriad of other undiscovered viruses that in some cases sit quietly inside hosts that are themselves unknown. Viruses are able to multiply only within the living cells of some other organism, usually an animal or a plant with which they have established an intimate, ancient and often (but not always) mutually beneficial relationship. In most cases, then, they are benevolent parasites, unable to live outside their host and not doing too much damage. Once in a while they kill a monkey or a bird here and there, but their carcasses are quickly metabolized by the jungle. Humans hardly ever notice.

Three. Today, however, the destruction of ecosystems seems to have as one of its consequences the increasingly frequent appearance of pathogens in areas larger than their original ones. Where trees are felled and fauna killed, local germs are found flying around like dust rising from rubble. A pest disturbed in its daily life and evicted from its usual host has two options: find a new home, a new kind of home, or become extinct. So, it's not us they're mad at, it's us who have become annoying, visible and very abundant. "If we look at the planet from the point of view of a hungry virus," writes historian William H. McNeill, "or of a bacterium, we see a wonderful banquet with billions of available human bodies, which until recently were about half of what they are now, because in twenty-five or twenty-seven years we have doubled in number. We are an excellent target for all those organisms that can adapt just enough to invade us". Viruses, especially those of a certain type, whose genome consists of RNA and not DNA and is therefore more subject to mutations, adapt well and quickly to new conditions [e.g. the recent (2021) Brasilian and Indian variants of COVID-19].

All these factors have led not only to the emergence of new diseases and isolated tragedies, but to new epidemics and pandemics, the most terrible, catastrophic and sadly notorious of which is that caused by a virus classified as HIV-1 group M (there are eleven other relatives), that is, the one that causes most cases of AIDS in the world. It has already killed thirty million people since its appearance some thirty years ago and today another thirty-four million or so are infected. Despite its planetary spread, few people know the fatal combination of events that led HIV-1 group M to emerge from the remote African jungle where its ancestors were ape hosts, seemingly without causing harm, and enter the course of human history. Very few know that the real story of AIDS does not begin among the American homosexual community in 1981 or in some African metropolis in the 1960s, but fifty years earlier at the headwaters of a river called the Sangha in the jungles of southeastern Cameroon. Even fewer have heard of the startling discoveries of recent years, which have enabled us to add detail to the story and to revise our positions".

I do not think it is necessary to add any more.

Already Konrad Lorenz, the famous ethologist, Nobel Prize winner in 1973, in his book "The eight deadly sins of our civilization" lists among others: overpopulation, the devastation of living space, competition among men, the deterioration of the genetic heritage. We are talking about fifty years ago!

But that is not enough. Twenty years ago, Hilary F. French, vice-president of the research sector of the World-watch Institute of Washington and, among other things, co-author of nine editions of the State of the World report, highlighted the same aspects. Just read some of the sentences in her book Vanishing Borders—Protecting the planet in the age of globalization, published in 2000 by the World-watch Institute of Washington

"Globalization is one of the driving processes behind unprecedented biological erosion. Trade in timber, minerals and other natural wealth is growing, and many of the most biologically thriving areas are threatened by the surge in international investment in resource exploitation". Globalization becomes an element of accelerated exploitation of the environment with much shorter timescales than those needed for the environment to form and regenerate.

"But such barriers [those between initially isolated ecosystems] are becoming increasingly permeable as peoples and organisms spread causing a fragmentation of environments that has negative effects and unpredictable consequences".

It also talks about "bioinvasion" connected with the diffusion of allochthonous species, which often determines the destruction of autochthonous species, with the deep change of the interactions these last ones had with the rest of the local ecosystem, with effects which may be disastrous. Due to the introduction of exotic species, in the year 2000, it was estimated that about 20% of the vertebrates were at risk of extinction, with an increase which reached the 50% in the USA.

Finally, we report a short passage found in the paragraph entitled "Microbes from across the border": *"The deterioration of the environment is also a powerful facilitating factor for the migration of microorganisms. According to the World Health Organization, "ecological changes have contributed in one way or another to the emergence of many (if not all)" new diseases. Often the problem is linked to processes of deforestation and agricultural conversion of land, which alter long-held balances between microorganisms and host species. In other cases they are due to human behaviour or poverty conditions..."*

Konrad Lorenz was a scientist. And who listens to a "big wig" who is all about his studies and lives in his own particular world? Hilary French should have at least deserved a minimum of real attention from policy makers, given the body she worked for. Now that the pandemic has also affected the so-called civilized world, it is only to be hoped that we understand that we have to act in depth on many common paradigms of the globalized society.

After about two years unfortunately, but not unexpectedly, I did not yet see any radical change (perhaps not even simple trends of change) of mankind behaviour and mentality both about politics and environment.

6.4   Finally

What is the vision of time to be adopted in this respect? The most useful one to understand and try to solve environmental problems? First of all, let us try to understand that we are not in the situation of those who, while waiting for a train, try to position themselves on the platform in the best way to easily find a place. The train has already left, and we are running to try to catch it anyway. The damage has already been done, and we only have the chance to mitigate the consequences and not do any more. It is no longer the time for preventive medicine, but for emergency medicine. We need competent "doctors" who, by making full use of their in-depth knowledge, are able to make quick and effective decisions. And, as it happens, especially in serious cases, we need a team work with close-knit elements who cooperate and not self-referential and inconclusive prima donnas who strut their stuff. Either we focus on the problem and collaborate to try to solve it or we waste time, Gaia's and our time, to point the Cassandras to public ridicule, forgetting, however, that Cassandra was right. There is only one choice; there is no place for "tinkering" or, worse, false illusions. Said Kenneth E. Boulding (1910–1993), economist born in Great Britain and naturalized in the USA: "*Anyone who believes that exponential growth can last forever within a bounded world is either a fool or an economist*".

If you were on a plane in trouble do you think you would argue about landing at the nearest airport to the destination you had chosen, or would you agree to get off as soon as possible? In the meantime let us make ourselves safe, then we will see either we will cancel the trip or we will resume it maybe by train, now it is not important.

Remember that we have often spoken of past, present and future. Perhaps we have not emphasized enough that although we cannot know the future, we can certainly affect it. If we have eaten a lot in the past, we are likely to gain weight. We do not know how much nor how our organism will react, but it is reasonable to foresee the increase of our weight and, perhaps, a fatigue in the functionality of the organism. During a long summer walk in the country-side, we brought a supply of water. Of course, it is logical to think that after a while we get thirsty. And we do not drink it all at once, but we ration it. Obviously, we have to dose it along the trip not to run out. Moreover, if we are with our children, we try to drink less of it ourselves to leave it to them. Do not you think this is all banal and logical? Certainly yes, but it seems that it is impossible to extrapolate these simple concepts and consequent actions to a larger stage.

I wonder if the ancient Greeks, in the current situation, could think of a future in which Gaia would do as the ancestor Kronos did when he devoured his children. Just as Kronos feared being ousted by those he had created so too might Gaia think of being ousted by human children.

I will gladly conclude this paragraph by referring again to Tiezzi, who in the book already cited says: "*Time, understood as the number of interconnected relations and as information stored in the energy-matter system, shapes the molecular forms in biological evolution. Time is not an abstraction, it is an integral part of matter, it is part of what exists and no political, social, economic theory can be described without taking into account the irreversibility of time itself. The problem of time is a fundamental problem, because the structures that connect us with each other and that are integral parts of the co-evolution of nature and human mind have time in them. …In other words, questioning the structure that connects humans with nature in a very complex history of biological evolution and reconstructing values and ways of thinking tending towards such integration, keeping in mind that we are 100% nature and 100% culture at the same time*".

References

1. Lovelock, J. (1991). *The new ages of Gaia*. Bollati Boringhieri (in Italian). Translated from: (1988). *The ages of Gaia—A biography of our living earth*. W.W. Norton &Company Inc. ©The Commonwealth Fund Program of Memorial Sloane-Kettering Cancer Center.
2. https://www.repubblica.it/ambiente/2016/10/02/news/james_lovelock_dieci_anni_fa_ero_certo_che_le_emissioni_di_CO2_e_il_global_warming_non_ci_avrebbero_dato_scampo_-148942414/. (Ten years ago I was sure that CO_2 emissions and global warming would not give us escape).
3. Tiezzi, E. (2001). *Historical times, biological times—Twenty years later* (p. 267). Donzelli Editore (in Italian).
4. Il Fatto Quotidiano. https://www.lifegate.it/persone/news/uomo-sterminato-83-per-cento-mammiferi-selvatici (in Italian). The original article is: Bar-Ona, Y. M., et al. (2018). The biomass distribution on Earth. *PNAS, 115*(2), 6506–6511.
5. Il Fatto Quotidiano. https://www.ilfattoquotidiano.it/2020/03/17/coronavirus-linquietante-terapia-del-gregge-di-boris-johnson-viene-da-lontano/5739091/ (in Italian).
6. Quammen, D. (2014). *Spillover—The evolution of pandemics*. Adelphi Ed. (in Italian). Translated from: (2012). *Spillover—Animal infections and the next human pandemic*. W.W. Norton & Company Inc.

Conclusions

Ask not, it is forbidden to know, what end the gods have appointed for me or for thee.

the gods have appointed, Oh Leuconoe, and do not consult the Babylonian Kabbalah. How much better, whatever it may be, to accept it!

Whether Jupiter has assigned several winters, or whether he has assigned as his last that which now wearies with the pumice cliffs that oppose the Tyrrhenian sea, be wise: filter the wine and to a short term limit the long hope.

As we speak it will have fled, inexorable, the time: **seize the day,** *as little as possible trusting the next.*

Do you recognize it? It is Horace, and, written in bold type, there is the translation of "carpe diem". It is an invitation to consciously and responsibly enjoy what you have at the moment, given the unpredictability of the future. It is, therefore, an exhortation to make the most of every moment of one's existence as one lives it, aware of how ephemeral it is. It does not mean to propose, as some think, a selfish attitude solely and blindly focused only on oneself and one's pleasures. Of course, to know what you have, you must first understand it and it is not easy. Time is part of it, and perhaps it is the most precious "commodity". I think I am able to understand its value, but I stop there.

I would be very happy if you could draw some conclusions about what time is. I do not, and I have no hope of it. I tend to be pragmatic and believe that it is convenient for us to take, from time to time, the point of view that best helps us to live reasonably both as individuals and as humanity and perhaps, to quote Lovelock, united with Gaia. The ways of thinking

W. Grassi, *The Challenges of Time*, https://doi.org/10.1007/978-3-030-94372-1

about time are various and depend on the perspective with which we observe it. Once more, the observed and the observer interfere. We have forced the cyclical repetition of the seasons down to a daily cycle. Nature imposes only sunrise and sunset on us, and we scan the hours and minutes of the daily routine. We have valued time according to the things we have to do, the results we have to achieve, the deadlines and so on. We come out with this from the circular repetition of things because the goals we set ourselves are intended to progress by gradually improving the conditions of life.

The point is that we do not know how much we want to improve them or how far we will go to do so. We risk a pointless race to no one knows what.

We add to the reassuring cyclicity proposed by nature, which, however, could seem monotonous and demotivating, a direction of time towards a future of individual progress. It is a way of accentuating the perception of time, but also of letting it pass with less attention. In any case, today's nine o'clock is not the same as yesterday's nine o'clock, nor is it the same as tomorrow's nine o'clock, but if you have perspectives and motivation you will feel the difference more strongly. How many times tomorrow was the day you would meet your girlfriend again. You were looking forward to the joy of the event and made plans to savour every moment of it. At the same time the time you were thinking about all this and the time between now and tomorrow you felt at most like a third wheel.

At the same time, we know the inexorability of the passing of time, as well as its capacity for improvisation, which in an instant can distort existence.

However, everything would change if we lived in other conditions. For example, if we had different sensory capabilities, or the events that affect us (light signals) occurred at very high speeds and/or we were very small in size. Think of you standing still in Houston talking to a friend who is on a space shuttle on its way to Mars. Let us assume that the shuttle has a speed of two hundred thousand miles per hour. Both of you would not be able to appreciate any relativistic effect (basically a factor γ of 1). Start the interview by calling him/her by name. For a short time, your conversation remains in a sort of limbo: you have already spoken, and he has not heard you yet. The friend's name is somewhere in your past (you have already pronounced it) according to the watch on your wrist. It has not reached him, it will shortly, in his future and in yours. And it is in your future that he will answer you, which for him is the present for the answer and so on throughout the interview. The more the friend moves away, the more these "dead times" increase.

The same if we were muons we would touch with our own hands that impossibility to understand nature in full, as Heisenberg spoke of.

But being as we are, elements of a very large, but finite and extremely complex macroscopic system, with a large number of interacting timescales, forces us to look far beyond our individual reality. Moreover, it puts us in front of a responsibility that Horace could not foresee: we do not know our future as individuals, but we can condition the future of our species and perhaps of the whole system of which we are part.

Beyond the various perspectives from which we look at this thing we call time and the possibility they give us of looking at it from different angles, I think it will continue to be a special observed on which it will be very difficult to pronounce a definitive word.

As human beings, we try to simply accept that this is how things are. What consequences can we draw from this? First of all, if time did not exist, as a succession of events, it would not make sense to refer to the past, the present and the future. A fortiori neither would it make sense to refer to what Saint Augustine calls the present of the past, the present of the present and the present of the future. So no more memories of what has happened or aspirations and hopes for what is yet to come. I do not remember my young son or even try to imagine what he will be like when he grows up. No memory or aspirations. It seems sad and surely inconceivable.

Thankfully, things are different and by "feeling the events flow" we are able to feel emotions, remember and make plans, for better or worse.

The uncertainty of the future, which is not of tomorrow but of now, should push us in the direction suggested by Horace. But we are not alone and what we feel does not depend only on us. What we experience is a function of our interactions with what surrounds us and how our sensitivity filters them. A sunset can seem to us a stupendous kaleidoscope of colours when we admire it at the end of a peaceful day. It can give us tranquillity and hope for tomorrow, accentuating the tenderness for the woman we are watching it with on the seashore. It gives us a sense of dismay and finality if we see it from a hospital bed before a dangerous surgery. But it gives us a sense of relaxation and the end of a nightmare (like "finally this is done too") after the surgery is happily over. The range of sensations caused by the same event is endless. The interaction with it changes depending on the succession of things that have occurred up to that moment (a clear day or not, before or after the operation), on the presence of another individual and on the relationship that binds you to him, and so on.

This is a sort of "personal time" that even though it is "ours" we do not determine by ourselves. We are "open systems" and how we live it is also due to the part of the outside with which we are in relation moment by moment. Not only this. We also condition the moments of those close to us. To put it in a more detached way, our moments are influenced by what is spatially (even optically) close to us. At the same time, our actions affect a more or less limited spatial scope. On this scale, however, they are irreversible and there is no going back to experience them in another way.

The scope in which one's actions are reflected increases disproportionately according to the power one has, especially if one exercises it with ignorance, conceit, arrogance and selfishness, as seems to me to be happening more and more frequently. Many, equally guilty, prefer to refrain from actively taking any position.

Let me make a comparison. You are driving on the highway, and you do not keep a safe distance from those in front of you, no problem the traffic is flowing. Your travel companions do not point it out to you, out of courtesy, because they trust your reflexes, because telling you something is useless or simply they do not care about your driving because they prefer, much more pleasantly, to photograph the landscape. On the other hand, you have always driven this way without any problem. Suddenly, there is a traffic jam and the truck in front of you brakes sharply. You have no braking distance, and you already know what happens.

Yet, you know that something can happen suddenly, and the safe distance is there to give you time to react. In nature, too, many things can happen abruptly. There are what are called threshold values. Once certain quantities exceed these values, the phenomena change abruptly over time and can become uncontrollable. The next time you make tea, look at what water does on fire. First it heats up and remains liquid; at most you can see some movement. Then, it starts to form little bubbles that turn into big bubbles of steam. It is telling you: "look, I can't stay in the liquid state anymore, I'm starting to turn into steam". Obviously for us, the message is different, and it invites us to turn off the gas and put the tea bag in. If we did not turn off the gas, which is the cause of the heating of the water, after a certain time we would be able to destroy the pot too. Gaia has already sent us signals and continues to send them as we continue to think we have time, continuing to shorten the safety distance.

Printed in the United States
by Baker & Taylor Publisher Services